THE TRANSITION HANDBOOK

"This book by the visionary architect of the Transition movement is a must-read, labelled 'immediate'. Growing numbers with their microscopes trained on peak oil are convinced that we have very little time to engineer resilience into our communities before the last energy crisis descends. This issue should be of urgent concern to every person who cares about their children, and all who hope there is a viable future for human civilisation post-petroleum."

– Jeremy Leggett, founder of Solarcentury and SolarAid, and author of
The Carbon War* and *Half Gone

"The Transition concept is one of the big ideas of our time. Peak oil and climate change can so often leave one feeling depressed and disempowered. What I love about the Transition approach is that it is inspirational, harnessing hope instead of guilt, and optimism instead of fear. *The Transition Handbook* will come to be seen as one of the seminal books which emerged at the end of the Oil Age and which offered a gentle helping hand in the transition to a more local, more human and ultimately more nourishing future."

– Patrick Holden, director of the Soil Association

"If ever there was a book that empowered the reader, this is it. I'm struggling here to escape metaphors about having a tankful of petrol in my belly, but that's just what it feels like. Rob tells us that fossil fuels multiply the physical force of each human being by 70 times. Well, this book can do the same, but in a social way rather than a brute mechanical way, and to a positive end rather than a destructive one. It's not only a powerful read, but an easy one too. It flows along like a well-written novel, full of illustrations, well designed and produced. Anyone who has met Rob or heard him speak in public will recognise in its words the humour, power and humility of this remarkable person. The book is of course a product of the cheap oil era. But if we can create things of this quality, when the post-peak times come we have little to fear."

– Patrick Whitefield, *Permaculture* Magazine

"The newly published *Transition Handbook* is so important that I am tempted just to confine this review to five simple words 'You must read this book!'."

– Richard Barnett, *Ethical Pulse*

"Rob Hopkins has written the most thorough description so far of how we get from the present chaos of cities and towns that are killing the planet and the people in them, to viable new ecologically sustainable urban and rural systems. This is more than a theoretical how-to manual; it is based on his own team's ground-breaking work, engaging whole communities in a transformative process that accepts the crucial need to reverse course, and has succeeded in doing so. The book is a great guide for how we must live in a future world where the limits of nature are honoured, but so are the basic comforts and joys of communities coming together in a great common cause. There is no more important book than this one for any community seeking change toward ecological sustainability."

– Jerry Mander, founder/director of the International Forum on Globalization
and author of *In the Absence of the Sacred*

"Rob Hopkins is the Gentle Giant of the green movement, and his timely and hugely important book reveals a fresh and empowering approach that will help us transition into a materially leaner but inwardly richer human experience. Full of reliable, readable, far-reaching scholarship, and warm-hearted practical advice on how to instigate transition culture wherever you are, this book will energise and regenerate your commitment to place, community and simple living. There is no better call to action than this book, and no better guide to the hands-on creation of a liveable future."

– Dr Stephan Harding, co-ordinator of the MSc in Holistic Science at Schumacher College
and author of *Animate Earth: Science, Intuition and Gaia*

"This DIY manual for change is an intelligent and practical attempt to encourage people to think globally while acting locally."

– P. D. Smith, *The Guardian*

"The Transition movement is the best news there's been for a long time, and this manual is a goldmine of inspiration to get you started."

– Phil England, *New Internationalist*

"If Hopkins is right about the viral spread of the Transition Town concept then he has to be a runaway contender for a Nobel prize."

– Friends of the Earth's *Earthmatters* magazine

THE TRANSITION HANDBOOK

From oil dependency to local resilience

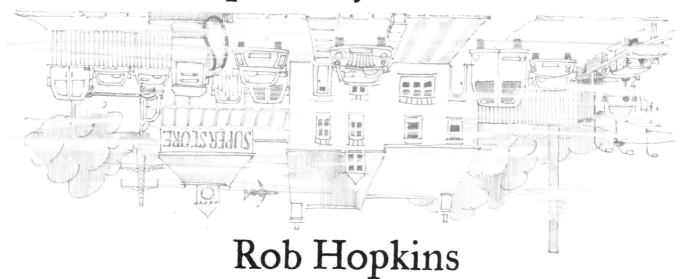

Rob Hopkins

CHELSEA GREEN PUBLISHING

WHITE RIVER JUNCTION, VERMONT

Contents

Contents

LIST OF FIGURES

FINDING 'TOOLS FOR TRANSITION' BOXES

Dedications

To my family: Mum, Dad, Jo, Ian, Jake, William, Steve, Hilary, Tessa, Robert, Harriet and Helen.

To Colin Campbell, Richard Heinberg, David Holmgren, David Fleming and Howard Odum for sowing the seeds of this concept.

To David Heath, Alan Langmaid, Barrington Weekes, Muriel Langford and Douglas Matthews for their insights into historic resilience.

To everyone testing the Transition model out, helping to nurture and water it and bring it to fruition.

To Geshe Jampa Gyatso, who taught me everything that really matters.

To Omeli, for a swift recovery.

To my wonderful sons, Rowan, Finn, Arlo and Cian; may they inherit this work's beautiful and abundant harvests.

To Emma: companion, lover, friend and now wife.

Originally published in 2008 by Green Books Ltd, Foxhole, Dartington, Totnes, Devon TQ9 6EB, UK.

First Chelsea Green printing November 2008. Printed in the United States of America

14 13 12   5 6 7 8

Our Commitment to Green Publishing
Chelsea Green sees publishing as a tool for cultural change and ecological stewardship. We strive to align our book manufacturing practices with our editorial mission and to reduce the impact of our business enterprise in the environment. We print our books and catalogs on chlorine-free recycled paper, using vegetable-based inks whenever possible. This book may cost slightly more because it was printed on paper that contains recycled fiber, and we hope you'll agree that it's worth it. Chelsea Green is a member of the Green Press Initiative (www.greenpressinitiative.org), a nonprofit coalition of publishers, manufacturers, and authors working to protect the world's endangered forests and conserve natural resources. **The Transition Handbook** was printed on FSC®-certified paper supplied by RR Donnelley that contains at least 30% postconsumer recycled fiber.

ISBN: 978-1-900322-18-8
Library of Congress Cataloging-in-Publication Data
Hopkins, Rob, 1968-
The transition handbook : from oil dependency to local resilience / Rob Hopkins ;
 foreword by Richard Heinberg.
p. cm.
Includes bibliographical references and index.
ISBN 978-1-900322-18-8
1. Energy consumption--Social aspects. 2. Energy policy--Citizen participation. 3. Energy conservation--Citizen participation. I.Title.
HD9502.A2H667 2008
333.79--dc22
2008043032

Chelsea Green Publishing Company
85 North Main Street, Suite 120, White River Junction, VT 05001
(802) 295-6300 www.chelseagreen.com

FSC
www.fsc.org
MIX
Paper from
responsible sources
FSC® C101537

Acknowledgements

Trailblazers and assorted sources of inspiration

Sharif Abdullah, Christopher Alexander, Peter Andrews, Peter Bane, Albert Bates, Graham Bell, David Boyle, Lester Brown, Colin Campbell, Fritjof Capra, Skye and Robin Clayfield, Alec Clifton-Taylor, Phil Corbett, Martin Crawford, Guy Dauncey, Josh Davis, Chris Day, Charles Dickens, Dr Carlo DiClemente, Chris Dixon, Richard Douthwaite, Matt Dunwell, Paul Ekins, Ianto Evans, Simon Fairlie, David Fleming, Elizabeth Fraser, Masanobu Fukuoka, Chellis Glendinning, Stephan Harding, Tim and Maddy Harland, Peter Harper, Lea Harrison, Robert Hart, Matt Haynes, Emilia Hazelip, Richard Heinberg, Colin Hines, Arthur Hollins, David Holmgren, Barbara Jones, Ken Jones, Martin Luther King, David Korten, Satish Kumar, Andy Langford, John Lane, Jeremy Leggett, Aldo Leopold, Bernard Lietaer, Jan Lundberg, Mark Lynas, Richard Mabey, Lucas Macfadden, Joanna Macy, Marcus McCabe, Dennis Meadows, Bill Mollison, George Monbiot, Helena Norberg-Hodge, Howard & Elisabeth Odum, Harrison Owen, Rosa Parks, Dr Stephen Rollnick, Mark Rudd, Kirkpatrick Sale, E. F. Schumacher, John Seymour, Vandana Shiva, Michael Shuman, Andrew Simms, Chris Skrebowski, Linda Smiley, Gary Snyder, Ruth Stout, Tom Vague and Meg Wheatley.

Collaborators, idea-shapers and co-conspirators

Paul Allen, Bart Anderson, Teresa Anderson, Sharon Astyk, Tom Atkins, Sophy Banks, Karen Blincoe, Ben Brangwyn, Lou Brown, Lynn Burke, Adrienne Campbell, Molly Scott Cato, Sky Chapman, Matthieu Daum, Catherine Dunne, Emma, Adam Fenderson, Naresh Giangrande, Brian Goodwin, Jennifer Gray, Mike Grenville, Pamela Grey, Colin Hines, Mike Jones, Marjana Kos, William Lana, Stephan Harding, Patrick Holden, David Johnson, Chris Johnstone, Tessa King, Kinsale Permaculture students past and present, Peter Lipman, Noel Longhurst, Caroline Lucas, Noni McKenzie, Rob Scot McLeod, Gavin Morris, Aaron Newton, Iain Oram, Dave Paul, Davie Philip, Adrian Porter, Tamzin Pinkerton, Hilary Prentice, Sarah Pugh, Thomas and Ulrike Riedmuller, Tom Rivett-Carnac, Louise Rooney, Mary-Jayne Rust, Chris Salisbury, Schumacher College, Simon Snowden, David Strahan, Graham Strouts, John Thuillier, Jill Tomalin, Nigel Topping, Totnes Sustainability Group, the people of Totnes, Robert Vint, Anne Ward and Patrick Whitefield.

Others without whom you wouldn't be reading this

The brave souls who read early drafts: Ben Brangwyn, Colin Campbell, David Fleming, Naresh Giangrande, Charles Hall, Stephan Harding, Tessa King, Peter Lipman, Mark Lynas, Rob Scot McLeod, Tamzin Pinkerton, Janet Richardson, David Strahan, Chris Vernon, people from the various Transition Initiatives and Carol Wellwood.

My wonderful editors, John Elford and Amanda Cuthbert at Green Books.

Louise Rooney and Catherine Dunne, who coined the term 'Transition Town'.

Photographers and illustrators

Thanks to Ron Hanna for the Captain Future images, Les Editions Albert René for the Asterix and Obelix picture, Tony Eriksen of The Oil Drum, *The Ecologist* magazine and Colin Campbell at ASPO for graphs, Dr Asgeir Sorteberg and National Snow and Ice Data Center for graphs, Andy Goldring for photos from Eastern Europe, The Imperial War Museum for Dr Carrot and other images, and Totnes Image Bank and Rural Archive for the use of their archive photos. Also Tulane Blyth, Adrienne Campbell, Stephen Gascoigne, Greenpeace/Ray Giguere, Nicholas Harvey, Sally Hewitt, istockphoto.com, Arthur Kay, Noel Longhurst, Nevenka Mulej, Rosalie Portman, Clare Richardson, Sally Stiles and Frankie Wellwood.

Foreword
by Richard Heinberg
author of *The Party's Over*, *Powerdown*,
Peak Everything and *The Oil Depletion Protocol*

I first encountered the Transition phenomenon in November 2006, when Rob Hopkins invited me to give an evening lecture in Totnes, Devon. Knowing that this was a fairly small town, I was hoping that fifty people might show up. Instead, over 400 packed the largest hall available. The same thing happened a few days later in Penzance, Cornwall, when Jennifer Gray invited me to kick off Transition Penwith; and again a few months after that at a similar event in Stroud.

Clearly a pattern was developing. The people at these events were not idly curious; they were champing at the bit to do something constructive in their communities about Peak Oil and Climate Change. I quickly formed the opinion that the Transition virus was the most exciting thing happening in the UK. On August 5, 2007, BBC Radio Scotland broadcast a story titled 'Towns Prepare for Peak Oil Point', which began with Rob calling the Transition efforts "one of the most dynamic and important social movements of the 21st century". The remainder of the radio programme provided plenty of evidence for that judgement.

Much of the Transition buzz can be traced to Rob Hopkins himself. A Permaculture teacher schooled in ecological design principles, he is by nature an intelligent, practical, sweet-tempered fellow, a family man with no apparent personal agenda other than ecosystem survival (he'd like his children to have a decent planet to live on).

In 2003 Rob was teaching in Kinsale, Ireland, when he first learned about Peak Oil directly from the world's premiere expert on the subject, petroleum geologist Colin Campbell. After sharing the information with his students, Rob worked with them to create the Kinsale Energy Descent Plan, which was later adopted as policy by the town council. It was the first strategic community planning document of its kind. Similar energy descent planning processes are now being undertaken in other towns and cities (including Portland, Oregon and Oakland, California), and in at least one industrial nation (Sweden).

Rob then decided to put on a Peak Oil conference in Kinsale in June, 2005 called 'Fuelling the Future', and that's where I first met him. After moving back to Britain to complete his doctorate, Rob decided to take the Peak Oil preparation process beyond the classroom by starting Transition Town Totnes in early 2006. That initiative took off like a rocket, and citizen groups in other towns throughout the UK quickly copied it.

Why is the Transition phenomenon so infectious? While there are efforts under way in well over a hundred communities around the world to deal with the looming implications of Peak Oil, there is something undeniably different about the Transition Towns – a sense of excitement, possibility, and engagement. Perhaps the buzz emanates partly from Rob's own contagious optimism. But this is no personality cult, since Hopkins is quick to cede the limelight to others whenever possible, and has designed the movement's governance process to be more 'bottom-up' than 'top-down'. To my mind, the best explanation is that Rob has hit upon a replicable strategy for harnessing the talents, vision, and goodwill of ordinary people.

And he has done so at a moment of extraordinary need.

There is simply no denying that we humans are facing tough times. Not only does evidence suggest that global oil production has already reached its all-time maximum and has begun its inevitable decline, but forecasts for natural gas extraction rates in the North Sea, North America, and Russia look worse than dismal. Meanwhile, new studies of global coal supplies suggest a peak in extraction rates could occur in as few as fifteen years, while the production of phosphates (essential for agriculture) is already down, as is global grain production per capita. The global climate is being destabilised, with Arctic ice melting faster than even the most dire scientific predictions said it would, while many countries are already experiencing a scarcity of fresh water. One could go on: if the 20th century was one of unprecedented growth in nearly every significant parameter (population, energy use, per capita consumption levels, etc.), the present century promises to be one characterised by declines in nearly all of those same categories, along with catastrophic weather events and drowning coastlines.

At the centre of the transition that the Transition Initiatives are engaged in is energy. Nearly all of the growth in population and consumption – as well as technological change – that occurred during the 20th century can be attributed to an unprecedented abundance of cheap energy, most of it from fossil fuels. Coal, oil and gas enabled the extraction and transformation of other natural resources at ever-accelerating rates, leading to the creation of enormous wealth amid widening circles of habitat destruction, pollution and climate chaos.

Fossil fuel depletion might be seen as a good thing, given the horrific environmental costs of using those fuels. But our societal dependencies on oil, coal, and gas constitute an enormous collective vulnerability, since there are no ready substitutes capable of fully replicating their services. Thus as fossil fuels go into decline, we will see a century of contraction in consumption levels that could cause the global economy to implode, undermining the survival prospects for the next generation. Unless we wean ourselves from these fuels proactively, societal support systems will crash just as the global climate gets pushed past a tipping point beyond which there will be nothing humans can do to avert worst-case impacts including sharply rising sea levels and devastated crops. Depletion and climate issues converge to make a deliberate, co-operative transition away from fossil fuels the centrepiece of our human survival strategy for the remainder of the 21st century.

On the whole, national governments are slow to understand and act on this imperative, as there are too many interests vested in maintaining the status quo. But if a country's leaders are largely unresponsive to the greatest crisis facing humanity, what's a concerned citizen to do?

The obvious answer: act locally. This especially makes sense in the present situation, because economic relocalisation will be one of the inevitable impacts of the end of cheap transportation fuels. We must produce more of our necessities close-by anyway; why not make the immediate community the source and focus of our entire energy transition strategy?

Rob Hopkins has grasped all of this in the Transition formula, and made it one that any community can enthusiastically buy into. Using Permaculture principles, the psychology of social marketing, and inclusive processes like 'open space', he has found a way for people worried about an environmental apocalypse to invest their efforts in ongoing collective action that ends up looking more like a party than a protest march.

This book is a 'how-to' guide to making it happen. It is like Rob himself—accessible, clear, and upbeat. If your town is not yet a Transition Town, here is guidance for making it one. If you are fortunate enough to live in a place that is already transitioning, you probably don't need my recommendation; you will have heard about this book though your web of personal connections.

In either case, make the most of it: we have little time and much to accomplish. And Rob Hopkins has offered us some invaluable tools for making our task easier and more enjoyable.

Richard Heinberg
Post Carbon Institute,
Santa Rosa, California

"It seems to me that [a] low-carbon society would be one which remembers that our planet is a unique gift – perhaps the only one of its kind in the entire universe – which we are indescribably privileged to be born into. It would be a society that could look back on the six degrees nightmare scenario as just that – a nightmare, one which humanity woke up from and avoided before it was too late. More than anything, it would be a society which survived and prospered, and which passed on this glorious inheritance – of ice caps, rainforests and thriving civilisations – to countless generations, far into the future."
– Mark Lynas

***

"When faced with a radical crisis, when the old way of being in the world, of interacting with each other and with the realm of nature doesn't work any more, when survival is threatened by seemingly insurmountable problems, an individual human – or a species – will either die or become extinct or rise above their limitations with an evolutionary leap. This is the state of humanity now, and this is its challenge."
– Eckhart Tolle

***

"Tell me, what it is that you plan to do with your one wild and precious life?"
– Mary Oliver

Introduction
Tantalising glimpses of resilience

Central to this book is the concept of resilience – familiar to ecologists, but less so to the rest of us. Resilience refers to the ability of a system, from individual people to whole economies, to hold together and maintain their ability to function in the face of change and shocks from the outside. This book, *The Transition Handbook*, argues that in our current (and long overdue) efforts to drastically cut carbon emissions, we must also give equal importance to the building, or more accurately to the rebuilding, of resilience. Indeed, I will argue that cutting emissions without resilience-building is ultimately futile. But what does resilience actually look like?

In 1990 I visited the Hunza Valley in northern Pakistan, which until the opening of the Karakorum Highway in 1978 had been almost completely cut off from the outside world. When I visited I knew nothing about permaculture, of the concept of resilience, or even a great deal about food, farming or the environment, but I knew when I arrived that this was an extraordinary place.

I found a quote in a book which I read as I travelled up towards Hunza (I no longer remember the title): "If on Earth there is a garden of bliss, it is this, it is this, it is this." They were words that replayed in my head many times over my two weeks in Hunza. Here was a society which lived within its limits and had evolved a dazzlingly sophisticated yet simple way of doing so. All the waste, including human waste, was carefully composted and returned to the land. The terraces which had been built into the mountainsides over centuries were irrigated through a network of channels that brought mineral-rich water from the glacier above down to the fields with astonishing precision.

Apricot trees were everywhere, as well as cherry, apple, almond and other fruit and nut trees. Around and beneath the trees grew potatoes, barley, wheat and other vegetables. The fields were orderly but not regimented. Plants grew in small blocks, rather than in huge monocultures. Being on the side of a mountain, I invariably had to walk up and down hills a great deal, and soon began to feel some of the fitness for which the people of Hunza are famed. The paths were lined with dry stone walls, and were designed for people and animals, not for cars.

People always seemed to have time to stop and talk to each other and spend time with the children who ran barefoot and dusty through the fields. Apricots were harvested and spread out to dry on the rooftops of the houses, a dazzling sight in the bright mountain sun. Buildings were built from locally-made mud bricks, warm in the winter and cool in the summer. And there was always the majestic splendour of the mountains towering above. Hunza is quite simply the most beautiful, tranquil, happy and abundant place I have ever visited, before or since.

At that time I was an artist, and spent my days with sketchbook in hand, wandering the fields, lanes and terraces, dazzled by the light and colour, spending many hours just working on one drawing in an ultimately futile attempt to try to represent the beauty of what was in front of me.

If (at that time) Hunza were to be cut off from the world and the global economy's highways of trucks packed with goods, it would have managed fine. If there were a global economic downturn, or even a collapse, it would have had little impact on the Hunza Valley. The people were resilient too, happy, healthy and with a strong sense of community.

I do not intend to romanticise or idealise it, but there was something I caught a glimpse of when I was in Hunza that resonated with a deep genetic memory somewhere within me. I grew up in England when the fossil fuel party was in full swing, in a culture ceaselessly trying to erase all traces of resilience and rubbishing the very idea at every opportunity, portraying country people as stupid, the traditional as 'old-fashioned' and growth and 'progress' as inevitable. In this remote valley I felt a yearning for something I couldn't quite put my finger on but which I now see as being resilience: a culture based on its ability to function indefinitely and to live within its limits, and able to thrive for having done so.

However, even then, in 1990, things were starting to change. When I was there, empty

Apricots drying on rooftops in Hunza. A drawing from the author's sketchbook, August 1990.

sacks of nitrogen fertiliser were visible in the corners of some of the fields. Sacks of cement were appearing, as were refined sugary foods and fizzy drinks. The process of undermining that resilience had begun in earnest, as has happened in most parts of the world and continues at a frantic pace. I haven't been back since, and so cannot offer an update, but I would be very surprised if the direction of change had been focused on the preservation of the Valley's ability to support itself. Indeed, from the amount of adverts on the internet for places selling 'Hunza produce', it appears to have moved towards being an export-driven economy.

Forces are converging very fast that make whether we choose to retain and enhance resilience, rather than just let it crumble, much more than just a philosophical discussion. It is no longer just a case of whether we should be questioning the forces of economic globalisation because they are unjust, inequitable or a rapacious destroyer of environments and cultures. Instead it is about looking at the Achilles heel of economic globalisation, one from which there is no protection other than resilience: its degree of oil dependency. The very notion of economic globalisation was only made possible by cheap liquid fossil fuels, and there is no adequate substitute for those on the scale we use them. The move towards more localised energy-efficient and productive living arrangements is not a choice; it is an inevitable direction for humanity.

The Transition Handbook is more than just a book of problems and ideas. It is about solutions, and about the Transition model, which I think may turn out to be the foundation for one of the most important social, political and cultural movements of the 21st century. I'd like to give you a brief taste of it.

It's a cool March evening in the small town of Totnes in Devon. Around 160 people are fill-

ing the seats of St John's Church for an evening event called 'Local Money, Local Skills, Local Power'. The event is run by Transition Town Totnes (TTT), the UK's first Transition Initiative, and the evening itself is something of an achievement: 160 people turning out to an event about economics, usually a subject guaranteed to stick people to their sofas tighter than superglue.

Each person, on arrival, is given a Totnes Pound, one of 300 notes produced by TTT as a pilot to see how a printed currency might be received in the town. One side is a facsimile of an 1810 Totnes banknote, from a time when Totnes banks issued their own currency, spotted four weeks before on the wall of a local filmmaker. As I begin my introduction to the evening and to the speaker, I invite the audience to each wave their Pounds in the air – it is quite a sight. 160 people, Pound in hand, beginning the powerful journey of telling new stories about money, and also about the future, its possibilities and their interdependence as a community.

The telling of stories is central to this book. You could think of it as being a story in itself: the story of the emergence of the Transition movement, of the most important research project taking place in the UK at the moment. It goes deeper than that, though. Our culture is underpinned by various stories, cultural myths that we all take for granted: that the future will be wealthier than the present, that economic growth can continue indefinitely, that we have become such an individualistic society that any common goals are unthinkable, that possessions can make you happy, and that economic globalisation is an inevitable process to which we have all given our consent. As we shall see, these are all stories that are profoundly misleading and indeed positively harmful for the challenges we find ourselves facing faster than we think. We need new stories that paint new possibilities, that reposition where we see ourselves in relation to the world around us, that entice us to view the changes ahead with anticipation of the possibilities they hold, and that will, ultimately, give us the strength to emerge at the other end into a new, but more nourishing, world.

As I stood at the front of that hall, watching the room full of laughing, twinkling people, waving their Totnes Pounds, I felt very moved. There is a power here, I thought, which has remained largely untapped. Surely when we think about peak oil and climate change we should feel horrified, afraid, overwhelmed? Yet here was a room full of people who were positively elated, yet were also looking the twin challenges of peak oil and climate change square in the face.

What might environmental campaigning look like if it strove to generate this sense of elation, rather than the guilt, anger and horror that most campaigning invokes? What might it look like if it strove to inspire, enthuse, and focus on possibilities rather than probabilities? We don't yet know for sure, but the Transition movement is an attempt to design abundant pathways down from the oil peak, to generate new stories about what might be waiting for us at the end of our descent, and to put resilience-building back at the heart of any plans we make for the future.

Transition Initiatives are not the only response to peak oil and climate change; any coherent national response will also need government and business responses at all levels. However, unless we can create this sense of anticipation, elation and a collective call to adventure on a wider scale, any government responses will be doomed to failure, or will need to battle protractedly against the will of the people. Imagine if there were a way of creating that sense of positive engagement and new storytelling on a settlement-wide, even a nationwide scale. This book is an exploration of that potential, an immersion in the possibilities of applied optimism, and an introduction to a movement growing so fast that by the time you read this book it will be larger still.

The time for seeing globalisation as an invincible and unassailable behemoth, or localisation as some kind of lifestyle choice, is over. The end of the Age of Cheap Oil is rapidly coming upon us, and life will radically change, whether we want it to or not. This book represents a new way of looking at what our future might hold, arguing that by taking a proactive response rather than a reactive one, we can still shape and form that future, within the rapidly changing energy context, in such a way that it ends up preferable to the present.

Rebuilding local agriculture and food production, localising energy production, rethinking healthcare, rediscovering local building materials in the context of zero energy building, rethinking how we manage waste, all build resilience and offer the potential of an extraordinary renaissance – economic, cultural and spiritual. I am not afraid of a world with less consumerism, less 'stuff' and no economic growth. Indeed, I am far more frightened of the opposite: that the process which took fertiliser sacks to the most fertile fields I will probably ever stand in continues, reducing the ability of communities to support themselves beyond the brief, transitory historical interlude when industry was able to turn natural gas into a fertiliser and when the car was king.

This is not a book about how dreadful the future could be; rather it is an invitation to join the hundreds of communities around the world who are taking the steps towards making a nourishing and abundant future a reality.

Rob Hopkins
Dartington, 2008

Part One

THE HEAD

Why peak oil and climate change mean that small is inevitable

"It is quite likely that the time interval before the global peak occurs will be briefer than the period required for societies to adapt themselves painlessly to a different energy regime."– Richard Heinberg[1]

"Any intelligent fool can make things bigger, more complex, and more violent. It takes a touch of genius – and a lot of courage – to move in the opposite direction." – Albert Einstein

"I feel it is my duty, given the social and economic chaos peak oil will undoubtedly produce, to stick very closely to defensible assumptions. If you ask me whether I personally think we'll make it to 2010, my answer is 'probably not'. Random factors and Murphy's Law more or less rule out everything running smoothly. This however is not analysis, but gut feel and hunch. On the hunch basis 2008 would be my answer, but 2010 my analysis."– Chris Skrebowski[2]

We live in momentous times: times when change is accelerating, and when the horror of what could happen if we do nothing and the brilliance of what we could achieve if we act can both, at times, be overwhelming. This book is underpinned by one simple premise: that the end of what we might call The Age of Cheap Oil (which lasted from 1859 until the present) is near at hand, and that for a society utterly dependent on it, this means enormous change; but that the future with less oil could be preferable to the present, if we plan sufficiently in advance with imagination and creativity.

This first part is called 'The Head' because it focuses on the concepts and issues central to the case that we need to be preparing for a future which looks very different from the present. It begins with an exploration of peak oil and climate change, the twin drivers of the Transition concept and the two greatest challenges facing humanity at the beginning of the 21st century (heading a long field of competitors). I will attempt to cover them in as

accessible a way as possible. It goes on to set out the nature of the challenges they present, and why they so urgently necessitate our rethinking a number of very basic assumptions as well as the scale at which we operate. Peak oil is dealt with first, and in more detail, because the likelihood is that you are less familiar with it. While climate change features widely in the media, peak oil has yet to register as a major issue, although the recent steep rises in prices are starting to change this. It is important to get up to speed on an issue of such central importance to our future.

I will go on to look at what kind of a world we could end up with if we don't respond imaginatively to these dual challenges, and then set out the thinking and the concepts underpinning Transition Initiatives. These initiatives are an emerging response: in essence, a powerful carbon reduction 'technology' and a new way of looking at responding to climate change and peak oil. They will be explored in depth as this book goes on.

Peak oil and climate change
The two great oversights of our times[3]

What is peak oil?: why it isn't the last drop that matters

There are plenty of other people better qualified than myself to tell you about peak oil.[4] I have never worked in the oil industry, am not a geologist, and other than having grown up in what is now one of the most rapidly depleting oil-producing nations in the world (the UK), I have no first-hand experience of oil production or geology. Prior to September 2004 I had never heard of the concept of peak oil, and had always assumed that oil in our economy worked in the same way as petrol in the tank of a car; that whether the engine was full or almost empty, it would run exactly the same. I thought we would potter along until some day in the distant future someone would put the very last drop of oil in their car and that would be that, a bit like the last truffula tree falling in Dr Seuss's *The Lorax*.[5] I was later to discover that I was somewhat wide of the mark, as I started to delve deeper into this incredibly important subject.

For me, learning about peak oil has been profoundly illuminating in terms of how I see the world and the way it works: the precarious nature of what we have come to see as how a society should function, as well as elements that any community responses we develop will need to have. Don't take my word for it – read around, inform yourself. Climate change – an issue of great severity – is only one half of the story; developing an understanding of peak oil is similarly essential. Together, these two issues have been referred to as the 'Hydrocarbon Twins'. They are so intertwined, that seen in isolation, a large part of the story remains untold.

Without cheap oil, you wouldn't be reading this book now. The centralised distribution of books would not have been feasible, and if you did have a copy, it would be one of only a very few books you had, and you would consider it a very precious possession indeed. I would not have been able to type it on my laptop, in a warm house, listening to CDs. When you really start thinking about it, it's not just this book that would not be here. Most things around you rely on cheap oil for their manufacture and transportation. Your furniture, entertainment, recreation, food, household appliances, medicines and cosmetics are all dependent on this miraculous material. This is not a criticism – it's just how it is for us all, and has been for as long as most of us can remember. It is almost impossible to imagine anything else.

It is entirely understandable how we got into this position. Oil is a remarkable substance. It was formed from prehistoric zooplankton and algae that covered the oceans 90-150 million years ago, ironically during two periods of global warming. It sank to the bottom of the ocean, was covered by sediment washed in from surrounding land, buried deeper and

> *"Sometime in 2006, mankind's thirst for oil will have crossed the milestone rate of 86 million barrels per day, which translates into a staggering 1,000 barrels a second! Picture an Olympic-sized swimming pool full of oil: we would drain it in about 15 seconds. In one day, we empty close to 5,500 such swimming pools."*
>
> *– Peter Tertzakian (2006),*
> *A Thousand Barrels a Second:*
> *the coming oil break point and*
> *the challenges facing an*
> *energy-dependent world,*
> *McGraw Hill*

deeper, and over time was heated under extreme pressure by geological processes, and eventually became oil.[6] Natural gas was formed through similar processes, but is formed more from vegetal remains or from oil that became 'overcooked' when buried too deep in the Earth's crust. One gallon of oil contains the equivalent of about 98 tons of the original surface-forming, algal matter, distilled over millennia, and which had itself collected enormous amounts of solar energy on the waves of the prehistoric ocean.[7] It is not for nothing that fossil fuels are sometimes referred to as 'ancient sunlight'. They are astonishingly energy-dense.

I like to think of fossil fuels being like the magic potion in Asterix and Obelix books. Goscinny and Uderzo's Gaulish heroes live in the only village to hold out against Roman occupation, thanks to a magic potion brewed to a secret recipe by their druid, Getafix. The potion gives them superhuman strength and makes them invincible, much to the chagrin of Julius Caesar. Like Asterix and Obelix's magic potion, oil makes us far stronger, faster and more productive than we have ever been, enabling our society to do between 70 and 100 times more work than would be possible without it.[8] We have lived with this potion for 150 years and, like Asterix and Obelix, have got used to thinking we will always have it, indeed we have designed our living arrangements in such a way as to be entirely dependent on it.

It is estimated that 40 litres of petrol in the tank of a car contains energy equivalent to 4 years human manual labour.[9] It is no wonder that we in the West consume on average about 16 barrels of oil a year per capita – less than Kuwait, where they use 36 (what do they do, bathe in it?), but far more than China's two, or India's less than one.[10] The amount of energy needed to maintain the average US citizen is the equivalent of 50 people on bicycles pedalling furiously in our back gardens day and night.[11] We have become dependent on these

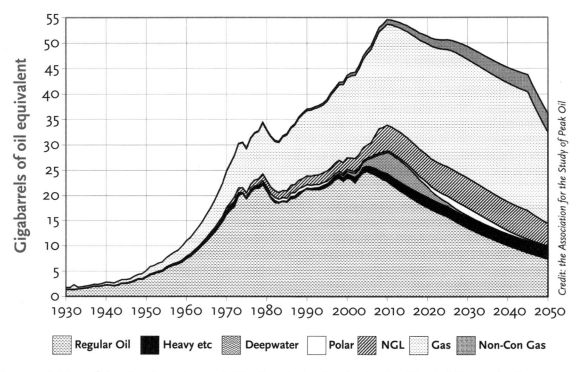

OIL & GAS PRODUCTION PROFILES
2006 Base Case

Credit: the Association for the Study of Peak Oil

Regular Oil   Heavy etc   Deepwater   Polar   NGL   Gas   Non-Con Gas

Figure 1. A Map of the Petroleum Interval. The Association for the Study of Peak Oil's graph of cumulative oil and gas production, which is one of the best researched maps of the possible future of oil and gas production.

"What is remarkable is the failure of politicians to start planning in any way for this inevitable transition, or even to start preparing their electorates for its inevitability."

– Jonathan Porritt (2007),
Capitalism as if the Earth Matters,
Earthscan

"Peak oil is a turning point in history of unparalleled magnitude, for never before has a resource as critical as oil become headed into decline from natural depletion without sight of a better substitute."

– Colin Campbell

pedallers – what some people refer to as 'energy slaves'.[12] But we are, it should also be acknowledged, extremely fortunate to live at a time in history with access to amounts of energy and a range of materials, products and possibilities that our ancestors couldn't even have imagined.

Figure 1 presents one of the best researched graphic representations of what we might call the 'The Petroleum Interval',[13] the brief interlude of 200 years where we extracted all of this amazing material from the ground and burnt it.

Viewed in the historical context of thousands of years, it is a brief spike. Viewed from where we stand now, it looks like the top of a mountain.

Oil has allowed us to create extraordinary technologies, cultures and discoveries, to set foot on the Moon and to perfect the Pop Tart. But can it go on forever? Of course not. Like any finite material, the faster we consume it, the faster it will be gone. We are like Asterix and Obelix realising, with a sinking feeling in the pit of the stomach, that the cauldron of potion they have in front of them is the last one. We can see the possibility of life without potion looming before us.

The key point here is that it is not the point when we use the last drop that matters. The moment that really matters is the peak, the moment when you realise that from that point onward there will always be less magic potion year-on-year, and that because of its increasing scarcity, it will become an increasingly expensive commodity. This year (2008), oil has for the first time broken through the $100 a barrel ceiling. Chris Skrebowski, editor of *Petroleum Review* magazine, defines peak oil thus, "the point when further expansion of oil production becomes impossible because new production

flows are fully offset by production declines".[14] It is the midway point – the moment when half of the reserves have been used up, sometimes referred to as 'peak oil' or the 'tipping point' that is important. It is a moment of historic importance. All the way up the slope towards the peak, since Drake drilled the first oil well in Pennsylvania in 1859, demand has driven supply. The more oil the world economy needed, the more the oil industry could produce.

'Swing producers' – that is, nations with large reserves which could increase output as required – ensured that supply could be increased whenever necessary. During the 1930s and 1940s it was the US that acted as that swing producer; in recent years it has been

"Fifty years ago, the world was consuming 4 billion barrels of oil per year and the average discovery was around 30 billion. Today we consume 30 billion barrels per year and the discovery rate is approaching 4 billion barrels of crude per year."

– Asia newspaper, 4 May 2005

"Energy experts no longer debate about whether Hubbert's peak will occur, but when".

– Fox News, April 28, 2006

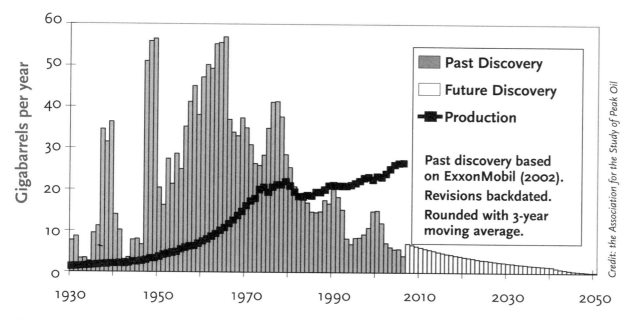

THE GROWING GAP
Regular Conventional Oil

Credit: the Association for the Study of Peak Oil

Figure 2. The peak in oil discovery and the widening gap between discovery and production.

Saudi Arabia. Once we pass the peak, supply begins to dictate demand, meaning that the prices start to rise suddenly and steeply, and the people with control of the remaining oil really get to start calling the shots.

Some key indications that we are nearing the peak

How might we know we are at or close to the peak? Firstly, there is an observable pattern that gives us an indication. Most oil-producing nations follow the same pattern – the peak in discovery tends to occur 30-40 years before a peak in production. Clearly one has to discover oil before one can produce it, and we tend to exploit the larger and easier reserves first. This pattern has been seen in the UK, the US, Russia and many more now-declining oil producing nations (see left). Given that the world as a whole peaked in discovery in 1965, we might, if the same pattern applies, imagine that we are close to, or at, the peak of production. This was first observed by geologist M. King Hubbert, who predicted in 1956 that the US would peak in production in 1970 (it had peaked in discovery in the 1930s). He was ridiculed, but eventually proved correct.[15]

Another indicator is that since January 2005, world oil production has stayed at between 84 and 87 million barrels a day (mbd), in spite of a very high price environment.[16] While the world economy desperately wants to increase consumption (the International Energy Agency has predicted that world production will reach 120mbd, a figure few in the industry take seriously), and oil prices have risen from $12 a barrel in 1988 to come close to $140 a barrel in June 2008. Its inability to keep up with burgeoning demand (see Figure 4, p.29) is a strong indication that matters geolog-

ical, rather than matters political or economic, are increasingly playing a role.[17]

Discoveries have fallen since their peak in 1965. This downward trend in discovery is also due to the fact that although we are still finding oil, the average size of the fields we are discovering is falling. In 1940 the average size of the fields found over the previous five years was 1.5 billion barrels, by 1960 it had fallen to 300 million barrels, by 2004 it was just 45 million barrels, and it continues to fall.[18] Indeed, during the Oil Age, 47,500 oil fields have been found, yet the 40 largest ones have yielded 75% of all the oil ever discovered.[19] As Figure 2 shows (page 21), the fall in discovery has been accompanied by rising consumption. 1981 was the year this gap began, and it has widened steadily ever since, to a point where we now consume about four barrels of oil for every one we discover.[20] In public, oil companies speak of high reserves and of a lucrative future. BP state that 'there is no reserves problem', Exxon that there is 'no sign of peak' and Aramco that there is 'no reserve problem'. Behind the scenes, however, they are increasingly aware of the nature of the problem. In November 2006, an event took place at Colorado Springs called the Hedberg Research Conference on Understanding World Oil Resources.[21] The event was by invitation only, and brought together people from across the oil industry, as well as from bodies like the United States Geological Survey, the International Energy Agency and the Energy Information Administration. No press were allowed, and people's presentations weren't shared. The day featured open and frank discussions, along the lines of "my company says this, but the data says this."

The intention of the conference was to reconcile the enormous difference in the estimates of likely future reserves additions between, on the

OIL-PRODUCING NATIONS THAT HAVE ALREADY PASSED THEIR PEAK

Albania, Argentina, Australia, Austria, Bahrain, Barbados, Belarus, Benin, Bulgaria, Cameroon, Chile, Colombia, Congo (Kinshasa), Croatia, Cuba, Czech Republic, Denmark, Egypt, France, Gabon, Georgia, Germany, Ghana, Greece, Hungary, Indonesia, Iran, Israel, Italy, Japan, Kyrgyzstan, Libya, Mexico, Morocco, Myanmar, Netherlands, New Zealand, Norway, Oman, Pakistan, Papua New Guinea, Peru, Poland, Romania, Russia, Senegal, Serbia & Montenegro, Slovakia, South Africa, Spain, Surinam, Syria, Taiwan, Tajikistan, Trinidad & Tobago, Tunisia, Turkey, Turkmenistan, Ukraine, United Kingdom, United States, Uzbekistan, Yemen.

– from www.EnergyFiles.com

one hand, what the US Geological Survey, creators of the most optimistic scenarios, produces; and, on the other, what other organisations believe to be the case. Companies brought along their detailed proprietary data, which is not made public, and tried to see if there was a clear pattern emerging. The results of this "I'll show you mine if you'll show me yours" session were striking. The USGS had put forward a figure of 650bn barrels yet to be discovered, but the conference put the figure at just 250bn. It also argued that the non-conventional oils (tar sands, deep water etc.) would never produce more than 4–5 million barrels a day, and indeed would struggle to achieve that, again much lower than the USGS. This kind of behind-the-scenes confidential meeting was also instrumental in the early days of climate change, leading to the formation of the Intergovernmental Panel on Climate Change.

A further indicator that we are nearly there is the nature of the new discoveries that the market gets excited about these days, and which are increasingly being expected to make up the shortfall as conventional oil production begins to decline. One of the new 'unconventional' sources of oil generating excitement is the Alberta tar sands in Canada. The problem with the tar sands is that the oil is very dense and viscous, more like sandy bitumen than oil. There are two ways the oil is extracted. The first is to dig it out with huge diggers, move it around in trucks the size of a house, and 'wash' the sands in the equivalent of a huge washing machine. Around 20% of it is produced this way. The rest is extracted *in situ*, where steam is pumped underground and the oil sucked out.[22] The resultant low-grade 'synfuel' is then refined into usable oil products. If the Alberta tar sands are the best we can do, we really are in trouble. Alberta is estimated to contain 175 billion barrels of oil, which makes Canada one of the top

four or five oil-producing countries in the world. Oil from tar sands is far more expensive to produce than most other sources of oil, but with the price of oil rising, these harder-to-extract oil sources become increasingly financially viable. Oil companies are moving into the area, and Fort McMurray, the area's main town, is becoming a boom town. Clive Mather, CEO of Shell Canada, describes Shell's operation in the area as the biggest thing he has ever seen the company undertake. People from all over the world are moving there for the 'New Gold Rush'.[23]

Tar sands production is somewhat akin to trying to remove the cocoa powder from a huge chocolate brownie. Greenpeace estimate that by 2011, annual carbon dioxide emissions from tar sands production will exceed 80 million tonnes of CO_2 equivalent, more than that currently emitted by all of Canada's cars.[24] Tar sands production also requires the felling of large areas of ancient boreal forest. The two principal weaknesses of the process are how the steam that separates the oil and sand is produced, and where the water to make that steam comes from.

You take clean, precious natural gas (a resource also on its own trajectory of depletion),[25] and burn it to make steam to produce 'synfuel', a poor quality dirty crude oil. It is madness. This is no 'gold rush'. Indeed, Matt Simmons, an energy industry investment banker, once described it thus: "Gentlemen, we have just turned gold into lead."[26] It is literally scraping the barrel, and rather than negating the peak oil argument – as those who say "look, see, there's loads left" propose – this confirms the peak oil argument: that we have reached the mid-point of the Oil Age, and the era of cheap oil is well and truly over.

One tyre alone, on one of the huge trucks, costs over £40,000. Tar sands production requires the price of oil to stay high in order to be

"Fuel is our economic lifeblood. The price of oil can be the difference between recession and recovery. The Western world is import-dependent."

– Tony Blair, April 7th 2002

"Clearly the timing of the global peak is crucial. If it were to happen soon, the consequences would be devastating. Oil has become the world's foremost energy resource. There is no ready substitute and decades will be required to wean societies from it. Peak oil could therefore constitute the greatest economic challenge since the dawn of the Industrial Revolution."

– Richard Heinberg

viable, but we should also ask how high does the price of natural gas have to rise before tar sands production becomes unviable again? The other limiting factor in tar sands production, alongside cheap natural gas, is water. It is estimated that it takes between two and four barrels of water for every barrel of synthetic crude produced from the tar sands.[27] The amount of water that can be extracted from the Athabasca River is finite, and is a major factor limiting production. Despite the lunacy of tar sands extraction, large amounts of money are pouring in to make it happen, due in part at least to the fact that it is one of very few places in the world open to private investment in oil production.

We might at this stage use the analogy of a pub. Conventional drilling of sweet crude oil, such as occurs in Saudi Arabia, is like standing at the bar while a charming barman pours you pints direct from the cask in the cellar. Tar sands are akin to arriving at the pub to find that all the beer is off, but so desperate are you for a drink that you begin to fantasise that in the thirty years this pub has been open for business, the equivalent of 5,000 pints have been spilt on this carpet, so you design a process whereby you boil up the carpet in order to extract the beer again. It is the desperate, futile action of an alcoholic unable to imagine life without the object of his addiction, and is only viable because oil prices are high and natural gas prices are cheap (high oil prices being the only one of the two that we can depend on).

Another recent story also indicates the less-than-reassuring nature of our new discoveries. It relates to a supposed huge find of new oil reserves in the Gulf of Mexico in 2006, "between 3 and 15 billion barrels" according to over-excited

Photo © Greenpeace/Ray Giguere

Oil production from the Alberta tar sands. The desperate scramble for 'unconventional oil' leads to degraded landscapes, soaring carbon emissions and a very low energy return on investment.

press speculation[28] (that's quite a range), which allowed many commentators to inform us that peak oil is now officially nonsense, and that we can all roll over and go back to sleep. This story was taken to show that there are still vast untapped reserves out there, that the peak oil 'doomsters' are wrong – and look, here in the Gulf of Mexico is the proof of that. When you read between the lines of this story, it isn't quite as exciting as it first appears (exciting that is, if you aren't someone who believes that the best place for oil is for it to stay in the ground). The oil, which could be bountiful or is more likely to be disappointing, is below one mile of ocean and four miles of rock, in the most hurricane-prone stretch of the Gulf of Mexico. Rental of the specialist rigs required to drill in such waters can cost half a million dollars *per day*.[29] Contrast this to the ease of drilling on land in the giant fields of Saudi Arabia or Mexico, and combine this with the falling size of new discoveries. . . it's clear that we are scrabbling around for crumbs.

The International Energy Agency, in a 2007

"If Iraqi production does not rise exponentially by 2015, we have a very big problem, even if Saudi Arabia fulfils all its promises. The numbers are very simple, there's no need to be an expert."

– Fatih Birol, Chief Economist, International Energy Agency, 2007

report,[30] talked about what it euphemistically called a 'supply crunch' in 2012. The reasons for this, they argued, are complex and diverse, but nothing to do with peak oil. Andrew Leonard at www.salon.com wrote:

> "To drastically summarise the report: The problem is not that the world is running out of oil, but that right now, offshore oil rigs are scarce and expensive, skilled labor is tight, transport infrastructure is limited, and political considerations such as 'resource nationalism' in states such as Venezuela and Russia and geopolitical risk in Iran and Nigeria are hampering investment and development. Logistics are the real problem, the report seems to be saying, and not the actual amount of oil in the ground. This leads to the conclusion that even though nearly 3 million barrels of new supply will be needed each year just to offset the decline in established oil fields, 'above-ground supply risks are seen exceeding below-ground risks in the medium term.'" [31]

However, Leonard is highly dubious of these given reasons, concluding "if it smells like peak oil, it probably is." Peak oil is the very large elephant in the room, one it is becoming increasingly difficult to ignore. Although all of the issues identified by the IEA are valid, they are increasingly being exacerbated by geological constraints.

The final reason that convinces me that we are close to the peak is the changing financial practices of the major oil companies. Firstly, the increasing and steadily more spectacular mergers between different oil companies, a practice sometimes referred to as 'prospecting on Wall Street'. An oil company's share price depends on its reserves, on the potential future production it has secured access to. As the trend in discoveries continues to fall, as it has done since 1965 (see Figure 2, p.21), it becomes harder and harder for companies to sustain their reserves

to offset against their production. It has become standard practice now for the larger oil companies to buy the smaller ones, thereby absorbing their reserves. Although oil companies have always done this, the scale of it has become increasingly dazzling.

A recent article by David Strahan examined the likelihood of a merger between BP and Shell, something that ordinarily would have been entirely unthinkable.[32] In spite of their protestations that peak oil is so far away as to be not worth thinking about, the move, if it goes ahead, would be primarily driven by the fact that they are producing oil but are increasingly unable to replace what they are producing with fresh reserves. BP managed to temporarily reverse its declines by engaging in the TNK–BP project in Russia, thereby adding the Russian company's reserves to its own, but now, shortly afterward, the gap is starting to open up again.

Another fascinating recent development has been large oil companies buying back their own shares. It is estimated that if Chevron Corporation keeps buying back its shares at current rates (it plans to spend $15 billion over the next three years), it will have liquidated all of its shares by 2023.[33] Exxon is doing the same, spending about $30 billion each year. With current high prices, oil companies are awash with money, but with few places to invest it. With discoveries falling, exploration is seen as yielding insufficient returns and providing a very poor investment. That peak oil is now a factor in the decisions of oil executives was spelled out in the '2007 Global Upstream Performance Review', which said:

> "We believe that the issue (peak oil) has become part of the industry's long-term planning. If the peak oil theory is correct, and a decline in world production is imminent, a company must choose among four alternatives – try to become

"Another reason why peak predictions are likely to be more accurate today is that areas of uncertainty are shrinking fast, a point made to me forcefully by Jason Nunn, a British director of the Washington-based consultancy PFC Energy. 'In the seventies and eighties when everybody said we were going to run out of oil, very few countries were in plateau, let alone decline. Today the majority of countries are in plateau, and many are in decline. Thirty years ago many parts of the world had not been explored, whereas today there are very few countries that haven't been. It's different now.'"

– David Strahan (2007),
The Last Oil Shock: a survival guide
to the imminent extinction
of petroleum man,
John Murray

Tools for Transition No. 1: The self-teaching peak oil talk

There are two ways one can deliver a peak oil presentation. The first is to stand up in front of a group of people and speak. The second is to get the group to give itself the presentation. How so, when they know nothing about it you might ask? Aha! This is where the Self-Teaching Peak Oil Talk comes in.

Simply print a good peak oil powerpoint (the Transition Network is producing a generic one which is a good place to start) onto A3 or A4 sheets, and on the back of each sheet put some text (short and to the point) which explains the slide.

Then distribute the cards among the participants and, like guests at a party, invite them to mingle and to tell each other the information on their 'slides'. As well as being a fun way to start getting people familiar with peak oil issues, it is also a great ice-breaker at an event or at the beginning of a course.

"Peak oil informs everything. People ought to know about that, but they don't. When it's going to peak or if it's happened already I don't know, but if oil ran out tomorrow we would be stuffed. We depend on it for everything."

– Zac Goldsmith, from Flintoff, J.P. (2007) 'You're going green. . . or else', *The Sunday Times*, September 9th 2007

a dominant participant, find a niche operational talent, harvest assets, or liquidate quickly." [34]

Share buybacks are a clear indicator of continued falling discoveries and return from investment in exploration, and suggest that oil companies are starting to plan for their own contraction.

These are, if you like, my top five reasons why peak oil is near. There are many more, as an investigation of the Resources section at the back of this book, or of some of the essential websites, will reveal. [35] At the end of the day, oil and gas are finite resources. It is clear now that at least 60 out of the 98 oil-producing nations of the world are in decline, and that even mighty oil-producing nations such as Saudi Arabia are experiencing enormous difficulties meeting demand. [36] Given that reaching peak oil will be a tipping point of unprecedented proportions, it seems reasonable then to ask: When might we expect to get there?

Peak When?

There is, as you might imagine, a wide range of predictions as to when exactly world oil production might peak, although recently this range has been narrowing. This diversity of opinion largely boils down to the fact that much of the information needed to make a precise prediction is not in the public arena. Around 80% of world oil is controlled by national oil companies, who have no obligation to make their reserves data public. In Saudi Arabia and Kuwait, for example, actual reserves data are a state secret and are fiercely protected. The private oil companies – the Shells and Totals of this world, responsible for a relatively small portion of the world's oil – are obliged to make their reserves data public. However, they have to walk a delicate line between keeping the regulatory authorities and their shareholders happy, and not revealing information of use to their competitors. Given the gravity of what peak oil would mean to the world, what we are left with is a decision: do we believe government and oil company assertions that all is well and that there is no cause for alarm in spite of the mounting evidence to the contrary, or do we question this complacency, and look more closely at what the oil industry is doing than what it is saying?

Environmental writer and campaigner George Monbiot puts it in stark terms: "Our hopes of a soft landing rest on just two propositions: that the oil producers figures are correct, and that governments act before they have to. I hope that reassures you." [37] In *The Upside of Down*, Thomas Homer-Dixon likens our situation to driving a car fast along a country lane in dense fog.[38] We know we are moving fast, we can hear the engine, but other than that it is hard to assess how fast we are moving. Our map suggests a straight road, and we are, after all, in a hurry. "Driving in fog is of course not sensible," he writes, "but it's exactly what we're doing today." Driving blind.

Kenneth Deffeyes, author of *Beyond Oil*,[39] publicly forecast his belief that peak oil for all oil would occur on Thanksgiving Day 2005 (that's Thursday 24th November for those of you outside the US).[40] Despite receiving much of the same ridicule dished out to M. King Hubbert, if we look only at conventional oil, he may well have been very nearly right. Conventional oil production appears to have peaked in May 2005 at 74.2 million barrels a day and has been declining ever since (see Figure 3).[41] Production of all liquids (that also

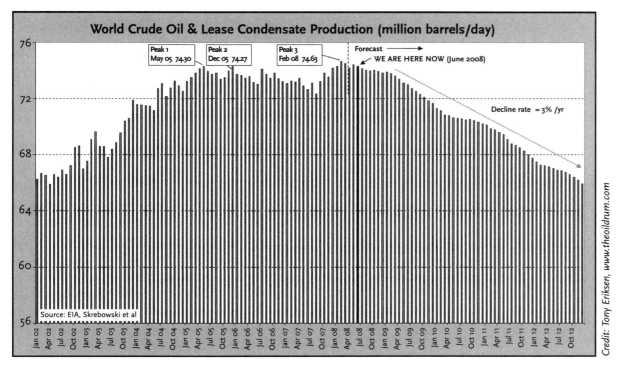

Figure 3. Production of conventional oil at least appears to have peaked in 2005, and the downward plateau has begun. The question is whether the unconventional sources of oil will be able to fill the emerging gap.

adds in tar sands, biofuels, deep water oil, all the harder-to-get-at stuff) has also plateaued over the last two years, despite sharply rising prices and enormous surges in demand from China and India, but this does not necessarily mean that we have actually peaked yet.

Other researchers give a range of dates. The Oil Depletion Analysis Centre puts peak oil in 2007,[42] Colin Campbell[43] and Chris Skrebowski[44] give 2010, and Jean Laherrere 2015.[45] The skeptics, such as Cambridge Energy Research Associates (CERA), no longer debate if oil will peak, rather when it will peak.[46] The CERA study, which generated a lot of 'peak oil theory is dead' media coverage when it was released in 2006, was thoroughly demolished (in my opinion) by an excellent piece by Dave Cohen, which analysed the CERA argument and found it lacking.[47]

Countering the report's claim that "CERA does not agree with the simplistic concept of an imminent peak in oil production nor with the idea that oil will 'run out' soon thereafter", Cohen wrote:

> "No one here or elsewhere is claiming that conventional oil will 'run out' anytime soon. Rather, the peak oil view is an evolving, sophisticated take on conventional oil production and the viability of substitutes to replace continuing demand for this paramount fossil fuel in the face of inevitable declines in available supply. Only the timing of such declines is at issue here. We can also only add that denial in the face of potentially very threatening events is a powerful force in the human psyche."

I am not qualified to give you an accurate prediction as to when this peak will occur. I am, however, by nature drawn to those who have no vested interests, who are independent of government or commercial interests, but who have analysed the data in depth. Predictions about

peak oil range from 'it has already happened' to 'it will never happen'. However, given that even some oil companies now acknowledge not only the concept, but have started putting dates to it, anyone who argues that there are 200 years' worth of oil left is living in cloud-cuckoo-land.

Thierry Desmarest, CEO of French oil company Total, recently said he thought world oil production would never exceed 100mbd,[48] telling a conference in Holland, "If we stay with this type of production growth, our impression is that peak oil could be reached around 2020."[49] Lord Ron Oxburgh, former Chairman of Shell, recently said world oil production "could well plateau within the next twenty years, and I guess I would be surprised if it hadn't." He added: "We may be sleepwalking into a problem which is actually going to be very serious and it may be too late to do anything about it by the time we are fully aware."[50]

In late October 2007, Germany's Energy Watch Group published a report which reassessed the data and argued very convincingly that world production had, in fact, already peaked in 2006, and "will start to decline at a rate of several percent per year". The report, which also argued that Middle East reserves are far lower than previously thought, concluded:

> "The world is at the beginning of a structural change of its economic system. This change will be triggered by declining fossil fuel supplies and will influence almost all aspects of our daily life. The now beginning transition period probably has its own rules which are valid only during this phase. Things might happen which we never experienced before and which we may never experience again once this transition period has ended. Our way of dealing with energy issues probably will have to change fundamentally."[51]

It is a great shame that the British Government

continues, in public at least, not to acknowledge the peak oil issue. A recent report on transport commissioned by the British Treasury stated that "fuel costs are forecast to fall by 26% up to 2025. An oil price of $35 a barrel is assumed in 2025." This forecast was published even though the price of oil was already at $50 a barrel by the time the report was released![52] More recently, in response to an online petition about peak oil, the UK Government wrote that "on the balance of the available analysis and evidence, the Government's assessment is that the world's oil and gas resources are sufficient to sustain economic growth for the foreseeable future."[53]

I think we can see Desmarest and Oxburgh's statements as the bookends in predictions of when the peak might occur. The majority of estimates are now falling between 2010 and 2015, with very few credible researchers placing their forecasts beyond this 2020 bookend. Having said that, the exact date of peak oil is really not so important. What matters is the fact that it is inevitable, it is going to be happening soon, and we haven't even begun to think what we might do about it.

How seeing the downward side of the mountain stretch away before us will affect our collective psyche remains to be seen. Figure 4 sums up our problem. We can see how closely supply and demand have followed each other, and how production has reached a plateau over the last two years. Once the peak is reached, though, the gap between supply and demand begins to steadily widen, and the price accordingly begins to rise sharply. It is often said that new ideas go through three stages. First they are

"The major oil companies have started making coded announcements indicating that they know the future of the oil business will not match its past. Instead of investing in production and discovery, all of the majors have been shedding staff and consolidating their holdings. None of this bespeaks a growing industry. And insiders know that there is very little excess capacity to be found anywhere."

– Dale Allen Pfeiffer (2006), Eating Fossil Fuels: oil, food and the coming crisis in agriculture, New Society Publishers

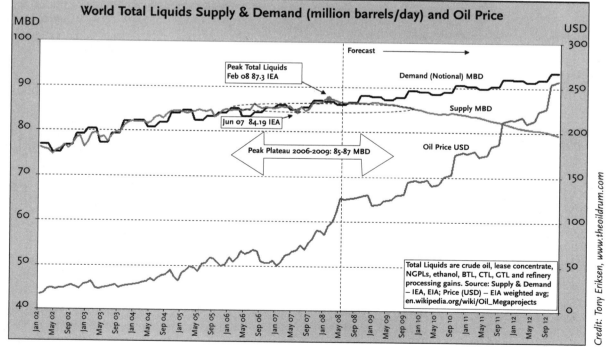

Credit: Tony Eriksen, www.theoildrum.com

Figure 4. Adding demand and rising oil price to the data from Figure 3, this graph highlights the impending gap opening up between supply and demand, what the IEA has euphemistically called a 'supply crunch'.

ridiculed, then they are ignored, and finally they are accepted as having always been the case. At the Association for the Study of Peak Oil conference in Cork, Ireland, in September 2007, former US Energy Secretary, James Schlesinger, said: "Conceptually the battle is over. The peakists have won. We're all peakists now."

Climate change

Until a year or so ago, climate change was seen as being such an unappealing subject to really embrace or get intimate with that most people felt happier looking the other way. Since then though, climate change has shifted much more towards the mainstream, with celebrities, governments and corporations falling over each other in the dash to become 'carbon neutral'. From the 'Live Earth' concerts to the Global Cool campaign which engages celebrities in raising awareness,[54] the campaign against climate change has grown rapidly. Supermarkets such as Tesco and Walmart are engaging in far-reaching analyses of their carbon footprints, and appear to be taking the issue very seriously, as do (in theory at least) politicians and policy makers. Even more than with peak oil, I write this section on climate change with great trepidation, as it is such a fast-moving field. Whatever I write will almost certainly have been overtaken by events by the time this book is printed. Climate change is happening faster than most models are able to keep up with, continually confounding expectation; models are being constantly revised and updated as the scale of this challenge becomes apparent.

When you start to explore the issue, climate change is extremely scary. Indeed, if it isn't scary, then you really haven't understood it. It is an area where one can easily resort to apocalyptic scare tactics, although I will try to avoid that

"We have at most ten years – not ten years to decide upon action, but ten years to alter fundamentally the trajectory of global greenhouse emissions."

– James Hansen, Director, NASA Goddard Institute for Space Studies

here; the information on its own is scary enough without dramatic embellishment. Sharon Astyk recently wrote: "One of the disturbing things about listening to scientists studying climate change is the fear in the voices and words of people not accustomed to be fearful, and the sense that generally speaking, scientists are far more worried than most of us are."[55]

We need to be realistic about where we are, and ambitious about what we can do. Climate change is a massive problem, but the worst effects could still be avoided if we are collectively able to engage with the issue. Transition Initiatives are but one of many powerful carbon-reduction technologies which, if embraced in time (and it is of course a big 'if'), can mean that we avoid the worst extremes of climate change. The trends at the moment, I grant you, really are not looking good.

The global climate is definitely warming; there is now no doubt about that. I don't need graphs, charts and scientific papers to convince me of that. Just within my own lifetime, I have seen the climate changing. I remember as a child winters being far colder, having to dig the snow away from the front doors, and power cuts caused by snowfall. Then, unsettled weather was the norm. Now, the weather is just plain unsettling, and, as we shall see, it will continue to become more so. Records are constantly being broken. In the UK, April 2007 was the hottest April on record, June 2007 the wettest June, autumn 2006 the hottest autumn, spring 2007 the hottest spring, July 2006 the hottest month, and the summer of 2007 was only a few millimetres away from being the wettest summer.[56] As someone on BBC Radio 4's *The Now Show* said recently, "I don't know about carbon emission levels but I do know that when a wasp lands on my Christmas cake something is not right."

On the astonishingly wet night of Friday, 20th July 2007 (the night when floods submerged large parts of the Midlands), I remember hearing on the radio that four times the average rainfall for July had fallen in two hours. At the same time, Greece was having unusually hot weather, leading to the dreadful forest fires that engulfed the country a month later, with vast plumes of smoke visible in satellite photos of the country. We all have observations from our daily lives of the climate changing, whether it is seeing flowers out far earlier than previously, swallows arriving a month earlier than usual, as they did in 2007, or the fact that we have to turn on our heating in the winter less often. In some cases, people go to extraordinary lengths to pretend it isn't happening (see daffodils article on the following page).

The greenhouse effect

The greenhouse effect isn't something we recently invented – without it, no life would exist on this planet. Without the layer of carbon

Photo © istockphoto.com

dioxide and other gases keeping the warmth in, our average global temperature would be −18°C. The Earth has gone through various ages of warming and cooling during its history. During the most recent Ice Age, 18,000 years ago, half of the UK was under a mile of ice, and so much water was locked up in the ice sheets that sea levels were up to 75 metres lower than at present. Scientists have recently found remnants of entire landscapes, with settlements, lakes, forests, marshes and hills under 450ft of water in what is now the North Sea. This discovery emerged from analysis of seismic data from oil companies, and has been christened by scientists 'Doggerland'.[57]

Professor Vince Gaffney of the Institute of Archaeology and Antiquity at Birmingham University, who led the research, said: "The coasts, rivers, marshes and hills we found were, for thousands of years, parts of a landscape that would have been familiar to hundreds of thousands of people and countless species of animals. Now it is all gone." So, although having the greenhouse effect has been one of the factors that has made life on Earth possible, the problem comes when the gases that form that layer (carbon dioxide, methane, nitrous oxide and so on) build up and trap more and more heat in the Earth's atmosphere. It is akin to throwing more and more duvets onto a bed, which leads to the problems discussed below.

CO_2 is such a small part of the overall atmosphere around us that it is measured in parts per million (ppm). Normally anything that is measured in parts per million is so insignificant that it is hardly worth bothering about. Pre-industrial levels of carbon were 278ppm, but by 2007 they have reached 385ppm. This seemingly trifling increase – caused by the incessant and ever-growing release into the atmosphere of carbon dioxide

"The scientific consensus presented in this comprehensive report about human-induced climate change should sound alarm bells in every national capital and in every local community."

– Klaus Topfer, United Nations Environment Program, commenting on the IPCC's Third Assessment Report, January 2001

A HOST OF PLASTIC DAFFODILS . . .
AN ODD MANIFESTATION OF CLIMATE CHANGE DENIAL

The Lake District is famous at this time of year for its amazing displays of daffodils, especially thanks to Wordsworth's poem, but the exceptionally warm winter and mild spring have meant that the daffodils have all flowered and wilted much earlier than usual, and, most importantly from an economic perspective, before the tourists arrive. The South Lakeland Parks holiday park at Fallbarrow, on the shores of Lake Windermere, has responded by planting thousands of plastic and silk ones instead.

Says spokesperson Caroline Guffogg:
"Our guests love to see the daffodils in bloom when they come for their Easter break, but this year the flowers have been out since the middle of February. The chances are they won't be at their best come April, so we've taken the decision to replace them. The fakes are high quality silk and are extremely realistic. Unless they look really hard then I don't think many people will notice the difference."

This strikes me as an extraordinary sign of the times. I wonder if the owners of the park have made the connection with climate change, and, as well as installing plastic daffs, have insulated their caravans, begun sourcing local food, started planting walnut trees and put solar panels up? If not, it is a demonstration of an amazing kind of denial, somewhat akin to men of a certain age covering their bald spot by sweeping their remaining hair over the top.

What next? Perhaps we should cover Mount Fuji with thousands of tons of fake snow just to keep the tourists happy. We could make some enormous plastic icebergs and tow them to the North Pole so we can pretend the real ones aren't melting. We could give some students summer jobs dressing up as orang-utans and swinging about in the trees in places where they are nearly extinct. Or we could just stop pretending and start responding. Really.

Transition Culture, Tuesday, March 20th 2007

from the combustion of fossil fuels, from changes in land use, from deforestation and so on, alongside the increases in emissions of methane from mining, livestock and the drying out of wetlands, as well as nitrous oxide from agriculture and aeroplanes – has already had significant effects, disrupting the delicate balance of the planetary climate. Although anthropogenic greenhouse gases (i.e. those caused by human activity) are only about 30% of total emissions, they have been enough to tip a very delicate balance.

The rise of carbon dioxide concentrations from 278ppm to 384ppm has led to global average temperature rising by 0.8°C above pre-industrial levels.[58] While this may not sound like much, just that level of increase has produced alarming changes around the world. These include widespread glacial retreat in the Himalayas, heavier than usual monsoons in India, Nepal and Bangladesh, encroaching drought in Australia, increasing frequency of tropical storms, as well as, it could be argued, the spectacular floods that hit the Midlands in the UK in the 'summer' of 2007. In Alaska, where average temperatures have increased 3–4°C, houses and roads are becoming unusable as the permafrost melts, which in turn releases more methane, a far more potent greenhouse gas than carbon dioxide. Sea levels are rising, and the rate of that rise is accelerating; the years 1993–2006 saw average rises of 3.3mm per year, far greater than the Intergovernmental Panel on Climate Change's prediction in 2001 of rises of 2mm per year.[59] Here in the UK, we are having to rethink the trees we plant, and the weather is noticeably changing. There is now no argument that the world is warming dangerously, possibly catastrophically, and there is an unprecedented scientific near-consensus that our oil-addicted lifestyles are to blame.[60]

So how high can we realistically allow world temperature to rise? The answer is that ideally we would stop all emissions today, but clearly that's not going to happen. As Mark Lynas describes so graphically in his book *Six Degrees*,[61] each degree that we allow world temperature to rise brings new and unprecedented scales of catastrophe. We still haven't broken through the 1°C threshold, but even so, the changes are clear to see. The ice-sheets on the Arctic Ocean are melting (of which more below), the Northern Passage was open to shipping in 2007 for the first time since records began,[62] droughts are increasing around the world, and weather records are being broken all the time. Hurricanes and typhoons are increasing, as are the number of heatwaves. Climate change is happening, and it's happening faster than the scientists' models can keep up with.

Is there such a thing as a safe limit?

If we break though the 1°C barrier, as now seems inevitable, we'll see a Mount Kilimanjaro completely bereft of ice, the almost complete collapse of the Great Barrier Reef, and a number of island nations submerged by rising sea levels. A 2°C rise would cause dreadful heatwaves, and increased drought around the world. Breaking through the 3°C barrier would mean that the growing season in Norway would be what it is in southern England today. The 3°C threshold would also bring about the complete collapse of the Amazon ecosystem, and the very real threat of conflict over water supplies around the world. Death rates from summer

heatwaves in Europe would make the summer of 2003 (which killed over 30,000 people) look positively tepid.[63]

Beyond that, in a nutshell, runaway climate change is not something you want to experience, or leave as a legacy to your children, yet we appear to be sailing alarmingly close to it. The emerging consensus in recent years has been that the imperative is to keep below 2°C at all costs. Even doing that, there is no guarantee that we will not have triggered runaway climate change. As George Monbiot puts it, two degrees is "merely less dangerous than what lies beyond",[64] and indeed a recent paper by James Hansen *et al.* at NASA argues that even 2°C is too high, given the rate of degeneration of the Arctic sea ice and the Greenland ice sheets, and that 1.5–1.7°C is more in line with adhering to the precautionary principle.[65] The reality is that the carbon dioxide already released will continue to push up the temperature for years to come (a phenomenon known as 'thermal inertia') by at least 0.6°C, meaning that we are already committed to a 1.4°C rise whatever we choose to do now. The warming we are experiencing now is the result of greenhouse gases emitted in the 1970s.

Let's return to the Arctic sea ice for a moment, as it may well turn out to hold the key to the future of human civilisation. The Intergovernmental Panel on Climate Change's Fourth Assessment Report in 2007 said: "Arctic sea ice is responding sensitively to global warming. While changes in winter sea ice cover are moderate, late summer sea ice is projected to disappear almost completely towards the end of the 21st century." [66] It appears, however, from a look at the increasing flow of literature on the subject,[67] that this melting is happening far faster than that, and that the ice is far more sensitive to rises in temperature than previously

"At just twenty-two weeks old, an average UK citizen will be responsible for the equivalent emissions of the greenhouse gas, carbon dioxide, which someone in Tanzania will generate in their whole lifetime."

– Andrew Simms *et al* (2006), *The UK Interdependence Report*, New Economics Foundation

thought. It has already reduced in size by 22% over the past two years, as well as becoming steadily thinner, halving in thickness since 2001.[68] Some predict now that the Arctic will be completely free of ice by 2013,[69] a hundred years ahead of the IPCC's forecast. This, in turn, will speed up the melting of the Greenland ice sheet, which is what could lead to sea level rises of as much as five metres by the end of the century, affecting two million square kilometres of low-lying land and 669 million people.[70]

Much of this re-evaluation is due to scientists only now beginning to develop models for the complex feedback mechanisms that influence rates of melting. James Hansen of NASA writes that "ice sheet disintegration starts slowly but multiple positive feedbacks can lead to rapid non-linear collapse."[71] While keeping below the 2°C threshold is vital, an increasing number of people are arguing that even 2°C is too little to prevent runaway climate change. David Spratt of Carbon Equity, having evaluated the latest evidence on the scale of the ice melting in the Arctic, writes that "[the IPCC's] most fundamental and widely supported tenet – that 2°C represents a reasonable maximum target if we are to avoid dangerous climate change – can no longer be defended." Given that we are not yet even at a 1°C rise, yet appear to have unleashed the catastrophic disintegration of the Arctic ice, 2°C is an absurd level to imagine as being 'safe' by any stretch of the imagination. He suggests that if we were able to wind back the clocks and start again, we would have based what constitutes 'safe' rises in emissions on

> "On the island where I live, it is possible to throw a stone from one side to the other. Our fears about sea level rise are very real. Our Cabinet has been exploring the possibility of buying land in a nearby country in case we become refugees of climate change."
>
> – Teleke Lauti, Minister for the Environment, Tuvalu

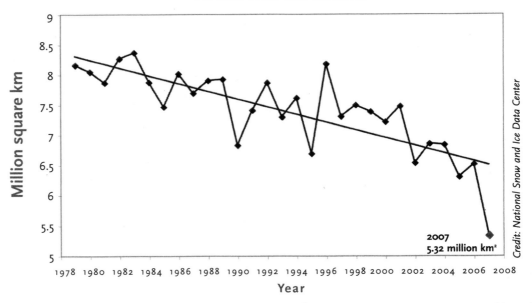

Arctic Sea Ice Extent

Credit: National Snow and Ice Data Center

Figure 5. A time-series of monthly average sea ice extent from August 1979 to August 2007. Note the precipitous fall-off of August 2007. The August 2007 monthly average extent was 5.32 million square kilometres, 31% below the long-term average of 7.67 million square kilometres. This sharp decline took those monitoring the Arctic ice by surprise.

what guaranteed the stable continuity of the Arctic ice sheet, which would probably have been around 0.5°C. The Industrial Revolution would have looked very different – or perhaps with the benefit of hindsight we would have decided to forgo it altogether.

Spratt concludes his study thus:

"The simple imperative is for us to very rapidly decarbonise the world economy and to put in place the means to draw down the existing excess CO2 levels. We must choose targets that can actually solve the problem in a timely way. It is not too late to be honest with ourselves and our fellow citizens."[72]

Cuts on this scale won't happen without an extraordinary, unprecedented, global concerted effort. This would be from a starting position where there is still no area of the world where

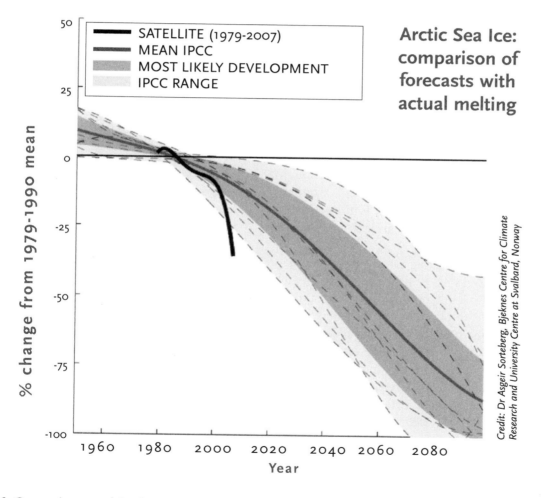

Figure 6. Comparing actual Arctic sea ice summer extent loss (up to September 2007) with IPCC projections. Note how the scale of melting has completely overtaken the IPCC's forecasts.

Credit: Dr Asgeir Sorteberg, Bjeknes Centre for Climate Research and University Centre at Svalbard, Norway

"If you think mitigated climate change is expensive, try unmitigated climate change."

– Dr Richard Gammon, University of Washington, on the steps of the US Congress, June 28, 1999

"In the last two years the analysis of climate dynamics has proceeded far beyond that portrayed in the latest IPCC Assessment Report. It was not taken into account in the Stern Report. The Climate Bill, currently before Parliament is based on even more out-of-date material, and is therefore utterly inadequate as a response to the current crisis. Acceleration of climate change is already a matter of observation. Virtually every parameter is moving faster than predicted by the international ensemble of climate models."

– David Wasdell (2007), in 'Planet Earth, We Have a Problem: feedback dynamics and the acceleration of climate change', All Party Parliamentary Climate Change Group

outputs of carbon emissions are actually falling. Until recently, it was believed that the scale of climate change necessitated cutting our emissions by 90% by 2050, or even by 2030, a mere twenty-two years away.[73] Trying to imagine maintaining our current lifestyles but emitting just 10% of the current amount of carbon is extremely difficult – almost unimaginable. However, buried in a report earlier in 2007 from the Intergovernmental Panel on Climate Change is an extraordinary piece of research. The IPCC researchers, using 'coupled modelling' (which basically means modelling including the impact of some feedback loops) concluded that to stay at under a 2°C increase in temperature, humanity has to zero its emissions by 2060.[74] That isn't saying that we have to achieve zero emissions from burning fossil fuels in our cars, planes and power stations (which might seem hard enough); that's zero from everything we do – from cutting down trees, from using fertiliser (manufactured from natural gas) and from raising livestock.

Similarly, reflecting on the implications of James Hansen's recent paper on Arctic ice melting, George Monbiot told the 2007 Climate Camp at Heathrow Airport: "We're not talking any more about measures which require a little bit of tweaking here and there, or a little bit of political tweaking here and there. We're talking about measures which require global revolutionary change." 90% cuts are no longer adequate, he said, nor, even, are 100% cuts. We are looking at 110-120% cuts, in other words sequestering more carbon than we produce. What it might actually look like – if you or I went to bed in the evening having sequestered more carbon than we had generated – I will consider later in the book, but this is clearly a monumental and unprecedented challenge.[75]

The intertwining of peak oil and climate change

One of the more absurd phenomena to emerge in recent years is that there are climate change activists who dismiss the peak oil argument, and peak oil activists who downplay climate change. It is as if people have discovered terrain which is somehow 'theirs', which they intend to gallantly defend against all-comers. I have spoken to a number of leading climate change activists who are doing great work on climate change, but who regularly want to downplay the peak oil issue. George Monbiot has expressed caution about emphasising the peak oil argument, fearing it will legitimise the case for biofuels, increased coal use, tar sands extraction and other climatically catastrophic approaches. "We don't have to invoke peak oil at all to see the sense and the logic in [the Transition approach], because even if the peak oil problem doesn't exist in any form, climate change does," [76] he told a public meeting in Lampeter.[77] However, in a subsequent article he revisited peak oil, examining the UK Government's predictions for increases in road transport, and asking what might power those cars, finding that, unbelievably "no report has ever been commissioned by the British Government on the issue of whether or not there is enough oil to sustain its transport programme." [78]

Tony Juniper of Friends of the Earth acknowledges that peak oil is a real challenge: "We do need to have the peak oil question in mind, because, irrespective of what we do about climate change, there is going to be an additional shock that's going to be economically significant, if not quite dangerous, coming from the oil price shooting up at some point, very likely in the not-too-distant future." He concludes, however, by saying: "So the two are related, but I think we have to keep them separate in terms of how we present them and deal with them because otherwise we create inadvertently damaging responses." [79]

I disagree. I will argue in this section that I don't think we can keep them separate, and that doing so does nothing to assist our development of realistic and potentially successful responses. Jeremy Leggett calls them the "Two Great Oversights of Our Times" [80] and, to borrow from Al Gore, peak oil is as much an Inconvenient Truth for climate change campaigners as climate change is for everyone else. Both, of course, are symptoms of a society hopelessly addicted to fossil fuels and the lifestyles they make possible. It is, however, too simplistic to assert that peak oil will mean climate change will be brought under control because we will run out of access to affordable liquid fuels; the situation is much more complex.

We do have a choice about how we respond to peak oil. We can use it as an argument for developing solutions that actually put in place infrastructure that will support us beyond the Oil Age, or we can use it to justify clinging to fossil fuels at all costs. The danger is, as Monbiot argues, that the gap which emerges as liquid fuels decline in availability will be filled with other fuels each far worse in terms of their climate impacts than oil was – the turning of coal into liquid fuels, tar sands, biodiesel and so on. If we don't fill the gap with conservation and a concerted programme of relocalisation (a concept explored in depth below), and if we refuse collectively to acknowledge the reality of energy descent (the downward trend in the net energy underpinning society), we will rapidly drive ourselves beyond the climatic tipping points and will unleash climate hell. If we see climate change as a separate and distinct issue from peak oil, we risk creating a world of lower

emissions but one which is, in terms of oil vulnerability, just as fragile as today's – if not more so – as energy prices rise.

A good example of this is New York, which recently emerged in a study as having one of the lowest per capita CO_2 emissions of any large Western city, less than a third of the per capita US average. [81] This is due to the density of living, the walkability, good public transport and the low heating requirements of apartment living. So, from a climate change perspective we can argue that New York is a good model of low carbon living we would all do well to emulate. Now let's weave peak oil into that mix. What happens to New York in the event of a power shortage, or when the price of importing food starts to rise sharply? New York experienced such a power cut in August 2003, and although it only lasted for a day, its impact was keenly felt. While New York may have a small carbon footprint, it has little or no resilience to declining oil supplies (a concept explored in depth in Chapter 3).

Climate change says we *should* change, whereas peak oil says we *will be forced to* change. Both categorically state that fossil fuels have no role to play in our future, and the sooner we can stop using them the better. It is key that both climate change and peak oil are given an equal degree of importance in any decision-making processes. It is interesting to observe that climate change is rapidly being taken on board by corporations, and increasingly by governments. Marks and Spencer now add labels to their clothes which say "If It's Not Dirty, Wash at 30," and supermarkets are falling over each other to be seen to be greener than their competitors. The idea of maintaining the global economy and just reducing its carbon output each year is attractive, and is now being seen as central to staying ahead of the competition. Apart from

"Taken together, Climate Change and Peak Oil make a nearly airtight argument. We should reduce our dependency on fossil fuels for the sake of future generations and the rest of the biosphere; but even if we choose not to do so because of the costs involved, the most important of those fossil fuels will soon become more scarce and expensive anyway, so complacency is simply not an option."

– Richard Heinberg

PEAK OIL

CLIMATE CHANGE

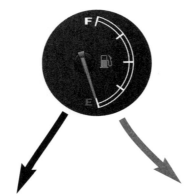

"Peak Oil and Climate Change are a bigger threat together than either are alone. Our biggest hope is to similarly converge our understanding of them, and how to deal with the problems they present. Peak Oil and Climate Change must be fused as issues – an approach is needed to deal with them as a package. If we are looking for answers, the environmental movement has pushed suitable ones for a long time. Peak Oil presents a tremendous chance to push those solutions ahead; failure to incorporate a full understanding of Peak Oil into the solutions argument for Climate Change would be an abject failure."

– James Howard,
Powerswitch.org.uk

PEAK OIL
(à la Hirsch Report)

- coal to liquids
- gas to liquids
- relaxed drilling regulations
- massively scaled biofuels
- tar sands and non-conventional oils
- resource nationalism and stockpiling

WHEN SEEN AS TWO ASPECTS OF THE SAME PROBLEM:

**BUILDING RESILIENCE
PLUS
CUTTING CARBON EMISSIONS**

Planned relocalisation
(building local resilience)

- tradable energy quotas
- decentralised energy infrastructure
- the Great Re-skilling
- localised food production (food feet)
- energy descent planning
- local currencies
- local medicinal capacity

CLIMATE CHANGE
(à la Stern Report)

- climate engineering
- carbon capture and storage
- tree-based carbon offsets
- international emissions trading
- climate adaptation
- improved transportation logistics
- nuclear power

Figure 7. What happens when you look at peak oil and climate change as two intertwined problems. (Adapted from Bryn Davidson: www.dynamiccities.squarespace.com)

the Swedish[82] and possibly the Irish[83] Governments, no government or corporation is yet really addressing or even acknowledging peak oil, at least publicly, because their business models will struggle greatly to adapt to it. For this reason the drive for reducing carbon emissions is coming largely from the top downwards; while responses to peak oil, due to its being less palatable to government and industry, appear to be coming more from the bottom up.

It is also important to point out that unless we plan in advance for peak oil, and adopt measures such as the Oil Depletion Protocol proposed by Colin Campbell and Richard Heinberg,[84] the recession caused by runaway oil prices will blow responses to climate change out of the water. Responding to climate change on an adequate scale requires a lot of money and an unprecedented degree of global co-operation. An economic recession – or worse, collapse – will make keeping the lights on our priority, and tackling climate change will slide rapidly down our list of priorities. Facing runaway climate change with a collapsed economy is the scenario we really want to avoid, and we separate these two issues at our peril. Perhaps one could also argue that while climate change offers globalised economies the possibility of gradual adaptation to a lower carbon way of continuing globalised international trade, peak oil asks much tougher, and possibly unanswerable, questions.

Figure 7 tries to set out what happens when peak oil and climate change are looked at together rather than in isolation. The Hirsch Report, which we will go on to explore, argues that we can mitigate peak oil with a crash programme of squeezing oil out of everything we can get our hands on. On the other hand, the Stern Report, commissioned by the UK Government to explore the economics of climate change, believes that climate mitigation and globalised economic growth are both possible and compatible. It does, however, completely ignore peak oil, stating that "there is enough fossil fuel in the ground to meet world consumption demand at reasonable cost until at least 2050",[85] an utterly absurd assertion in the light of what we have looked at in Chapter 1. I argue here, as Figure 7 shows, that when the two are combined, our options look very different, that when we combine the two, the rebuilding of resilience (a concept we shall go on to explore) is as important as cutting carbon emissions.

Can peak oil engage people more effectively than climate change?

Here I enter the realm of the contentious – and I do so cautiously but determinedly. It has been my experience from my work promoting the Transition concept that peak oil, if presented in the way that this book will go on to elucidate, can do more to engage and involve people and communities than climate change. Peak oil educator Richard Heinberg uses the analogy of a car: "At the most superficial level, we could say that climate change is an end-of-tailpipe problem, while peak oil is an into-fuel-tank problem."[86] One could add to this that people perceive themselves as being inherently more affected by rises in the price of a key commodity such as liquid fuels than by changes to the climate.

One of the things peak oil does very effectively is put a mirror up to a community and ask: "What has happened to the ability of this community to provide for its basic needs?" Allowing people to mentally explore what their current lifestyles would be like if the inflow of cheap oil were to cease is a powerful way to get people to think about the vulnerability of their oil-dependent state. It can focus the mind more than climate change because it can seem to be more obviously relevant to people's everyday lives. Also, for some, barrels of oil are easier to visualise than tonnes of gas.

If the 'early-toppers' such as Campbell and Skrebowski are right, peak oil will begin to visibly impact our lives within a few years. The impacts of climate change are still seen by many, despite the extraordinary flooding of the summer of 2007 and the accelerating collapse of the Arctic ice, as a more gradually unfolding process, while those of peak oil may be much more immediate.

Climate change is, rightly or wrongly, seen as a problem that will primarily affect the developing world before it affects the developed world, which, ironically, is largely responsible for creating the problem in the first place. The same is true, initially at least, for peak oil. While we in the West can theorise about what the impacts of peak oil might be, for many developing countries it is already a grim reality. Indeed, their enforced reduced consumption, sometimes termed 'demand destruction', could be seen as reducing consumption globally, thereby stopping runaway prices for those of us who can still afford liquid fuels.

Most countries in Africa, Asia and South America are already experiencing the effects of peak oil. Argentina is facing its worst energy shortage for twenty years, with widespread power cuts and natural gas shortages affecting public transport. Power cuts in Pakistan have led to riots, and in Iraq some provincial officials have begun disconnecting power stations from the national grid so as to keep the energy generated to themselves. Iran has introduced petrol rationing, and the UN recently warned the Sri Lankan Government that they will be unable to continue their humanitarian work in the country due to fuel shortages. In Uganda, grid power shortages have shut down the pipeline which brings diesel into the country from Kenya, a kind of peak oil 'feedback loop'.

In Nigeria only 19 out of 79 power plants work, and blackouts cost the economy $1 billion a year. Nicaragua is now running at a national energy deficit of 20-30%, with the national energy company having to shut down whole cities for 6-10 hours at a time. Costa Rica has regular blackouts, as does the Dominican Republic, where blackouts which originally only affected the poor *barrio* districts now extend to the exclusive residential districts.[87]

Thus far, the wealthier nations have been able to keep peak oil at arm's length thanks to an economic cushion which insulates us from rising prices, but only up to a certain point. Where exactly that point lies is hard to assess and nobody know for sure, other than to say that, if the price of oil rises above $102 a barrel, it will break the previous (adjusted for inflation) figure that caused a recession in the 1970s. Beyond $102 a barrel we are into uncharted territory, but its impacts on the economy are unlikely to be beneficial. Peak oil and climate change must be seen as equally pressing drivers for change.

The contradictions of the Hirsch Report

When the US Department of Energy commissioned Robert Hirsch and his colleagues to write a report looking at mitigation strategies for peak oil,[88] little could they have suspected what they were going to get. It also seems that Hirsch himself was unprepared for where the report would take him, and what he would end up writing. In an interview following the report's release he mused upon the report's conclusions:

"There's no question in my mind at least that peaking is likely to occur in maybe the next 10 or 15 years. So if depletion is as high as some people think it could be, we're in a very serious, serious problem. Much worse than the worst that we could think of. This problem is truly frightening. This problem is like nothing that I have ever seen in my lifetime and the more you think about it and the more you look at the numbers, the more uneasy any observer gets.

It's so easy to sound alarmist, and I fear that part of what I'm saying may sound alarmist, but there simply is no question that the risks here are beyond anything that any of us have ever dealt with.

The risks to our economies and our civilisation are enormous, and people don't want to hear that. I don't want to think about that. That's a very uncomfortable thing to think about. And I will tell you that it took some time after that realisation set in to be able to emerge and try to be positive and constructive about this problem. This is a really incredibly difficult, and incredibly severe problem." [89]

"A really incredibly difficult and incredibly severe problem" – clearly not a man to mince his words. Short of just scrawling "Aaargh!", this was as close as one could get to really conveying the immensity of the challenge of peak oil. The Hirsch Report was dynamite, and is seen by many in the peak oil community as being a seminal document, the first 'official' document to really take peak oil seriously. However, it is also worthy of deeper inspection, as it is not just a document about peak oil but it also offers an illuminating and terrifying insight into the responses to the challenge of peak oil that our leaders might pursue in order to keep the lights on and the wheels turning, if we fail to come up with anything better.

The nub of the report's problems can be summarised in the term "viable mitigation options". This term appears in the oft-quoted paragraph from the executive summary:

"The peaking of world oil production presents the US and the world with an unprecedented risk management problem. As peaking is approached, liquid fuel prices and price volatility will increase dramatically, and without timely mitigation, the economic, social and political costs will be unprecedented. Viable mitigation options exist on both the supply and demand sides, but to have substantial impact, they must be initiated more than a decade in advance of peaking."

The report's definition of what these options might be are profoundly at odds with what this book will propose. For Hirsch, viable mitigation options are sought from the basic premise that the show must go on in its current form, that business as usual must be preserved at all costs. In an interview I did with Richard Heinberg, I asked him what he saw as being the report's limitations.

"The implied goal [of the report] is to keep business as usual going as long as possible by any means necessary, including using coal to make liquid fuel. Of course, if it were feasible on any large scale, this would produce a climate catastrophe, but that was completely overlooked in the report. There's no evidence of concern for climate change issues whatsoever in the report. The goal of the authors is to suggest how we could keep the engines of modernity running as long as possible." [90]

In 2006, at the Association for the Study of Peak Oil conference in Italy, I heard Robert Hirsch give a talk called 'Mitigation of Peak Oil: Making the Case: more numbers and some

THE KEY FINDINGS OF
THE HIRSCH REPORT

• Although the exact date is uncertain, peak oil itself is an inevitability.

• The economic impacts of peak oil will be severe.

• Peak oil is a unique problem, both its scale and nature are unprecedented.

• The primary and most immediate problem is liquid fuels, for which replacements are hardest to find.

• Mitigation efforts will require substantial time, ideally 20 years, at least a decade.

• Both supply and demand must be addressed.

• Preparing for peak oil is a challenge of risk management.

• Government intervention will be required, but economic upheaval is not inevitable.

questions', which built on the 2005 report and set out a 'crash programme' to keep all the cars in the US on the road. His plan, he told the assembled audience, would cost $1 trillion a year, and would involve a massive expansion of coal-to-liquids, extraction from tar sands, gas-to-liquids and so on. I appeared to be one of only a handful of delegates looking utterly horrified at this proposal and thinking I must have misunderstood what he had said. I hadn't.

Musing aloud later on Transition Culture about the implications of Hirsch's talk, I wrote:

"Hirsch presented clearly what happens when one takes a purely peak oil perspective without the integration of a climate change one. For me, Hirsch laid out a clear and perfectly reasoned argument why we cannot possibly keep all our cars going and why we need to break our addiction to the car. He just hadn't realised that that was what he was doing." [91]

Imagine if the readers of this book were given a $1 trillion a year budget to initiate and drive a programme of global powerdown – think what could be achieved! There was some very dangerous thinking and there were some equally dangerous basic assumptions in Hirsch's presentation. I would not wish to take away from the usefulness of the Hirsch Report and his work on depletion profiles, but the recommendations in his talk, in the wrong hands, could lead to policy choices being taken which are in effect collective suicide.

The same really applies to the original Hirsch Report. If your starting assumption is that the show must go on at all costs, you will scrabble around for whatever strategies and technologies might, in theory, allow you to do so. The 'crash programme' advocated by Hirsch would greatly hasten the headlong plunge towards climate chaos.

Both the Hirsch Report and the Stern Report, as seen in Figure 7, illustrate the perils of looking at these two issues in isolation. Alternatively, when peak oil and climate change are seen as inseparable, we need to completely rethink our 'viable mitigation options', as well as acknowledge that business-as-usual is untenable. What 'viable mitigation options' might look like when the two are brought together will be explored as this book progresses, but the Hirsch Report offers us a clear exposition of what they absolutely must not be.

The other key aspect of the Hirsch Report is its assessment of timing. We will need, it argues, "more than a decade in advance of peaking" to prepare the economy for this transition, preferably twenty years. While this is a sobering way of looking at the scale of the challenge, I think that once a society decides to move, things can happen very quickly. Lester Brown cites the example of how the US economy re-geared itself entirely at the beginning of World War II. President Roosevelt, having set ambitious arms production goals, said: "Let no man say it cannot be done." The greatest expansion in output was in 1942, with the production and sale of private cars being banned, house-building and road construction halted, and driving for any non-essential purpose prohibited. Brown writes:

"The automobile industry went from producing nearly 4 million cars in 1941 to producing 24,000 tanks and 17,000 armoured cars in 1942 – but only 223,000 cars, and most of them were produced early in the year, before the conversion began. Essentially the auto industry was closed down in 1942 through the end of 1944. In 1940, the United States produced some 4,000 aircraft. In 1942, it produced 48,000. By the end of the war, more than 5,000 ships were added to the 1,000 that made up the American Merchant Fleet in 1939."[92]

When society decides to put its weight behind change, things can move very fast. A number of relatively small changes in legislation, giving people more money for energy from microgeneration than they'd pay the grid, carbon rationing, and changes to planning (i.e. promoting local agriculture and co-housing) will accelerate things greatly. While some of this needs to be driven at a national government level, much of the momentum and pressure, as well as the diversity of projects and initiatives that need sanction or support from government, can come from the local level. People need to hunger for these changes, and to see them as infinitely more desirable than the present.

While peak oil is a crucial insight into what is ahead of us, it is also essential to be mindful of some unsavoury proposals being sneaked in on the back of it. Just as climate change is sometimes put forward as the justification for expanded nuclear power and the illusory hydrogen economy, so peak oil (as the Hirsch report so graphically illustrates) can be used to instil a fear that we urgently need liquid fuels from wherever we can squeeze them. Some argue that it will usher in a Golden Age of coal. It is important to remain alert to that. As we will see in Part Two, this is a crisis which is about much more than what we'll put in our petrol tanks.

The lesson from the Hirsch Report, then, is that in proposing solutions to climate change and peak oil we must always be sure that we are asking the right questions. The question is not "How can we keep everything going as it is?" We should instead ask how we can learn to live within realistic energy constraints. Rather than deciding our plan of action first and then picking the energy options to match it, we should start by basing our choices on asking the right questions about the energy available to underpin our plans.

The Hirsch Report fails to ask the right questions. When devising solutions, we must address both climate change and peak oil from the outset. The 'viable mitigation options' we come up with will depend entirely on the nature of the questions we ask. Hopefully, this examination of the Hirsch Report has been useful in establishing why just looking at peak oil or climate change in isolation is both futile and potentially dangerous.

"Ladies and gentlemen, I have the answer! Incredible as it might seem, I have stumbled across the single technology which will save us from runaway climate change! From the goodness of my heart I offer it to you for free. No patents, no small print, no hidden clauses. Already this technology, a radical new kind of carbon capture and storage, is causing a stir among scientists. It is cheap, it is efficient and it can be deployed straight away. It is called . . . leaving fossil fuels in the ground."

– George Monbiot, 'Rigged', *The Guardian*, December 11th, 2007

Chapter 2

The view from the mountain-top

"One does not discover new lands without consenting to lose sight of the shore for a very long time."

– André Gide

I once had a conversation with an elderly man I met in a pub, which I thought was about peak oil. I talked about my understanding of it and he interrupted occasionally with fairly innocuous interjections, and I thought the conversation was getting somewhere interesting. At the end though, he said, by way of a closing flourish, "Yes, I used some of that on a table once – came up lovely." By this stage in the book I hope that you, unlike my elderly friend, will be feeling relatively up to speed with the nature of the principal challenges we face. You will be aware by now that our fossil-fuel-saturated lives will undergo a radical redesign whether we want them to or not, but you may quite reasonably be wondering what they might look like at the end of this redesign process.

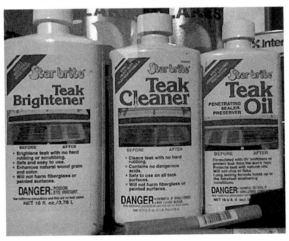

Evaluating possible ways forward

I do not have a crystal ball. I don't know how the twin crises of peak oil and climate change will unfold – nobody does. I don't know the exact date of peak oil, and again, nobody does. Similarly, I don't know if and when we will exceed the 2°C climate threshold, and what will happen if we do.

What I am certain of is that we are going to see extraordinary levels of change in every aspect of our lives. Indeed we have to see extraordinary levels of change if we are to navigate our societies away from dependence on cheap oil in such a way that they will be able to retain their social and ecological coherence and stability, and also live in a world with a relatively stable climate. In terms of looking forward, many people have set out different scenarios for what the future might hold. I have trawled through a lot of these for insights as to how life beyond the peak might be.

What I have set out to do in Figure 8 (p.46) is to position these very varied scenarios in relation to each other, starting at one end with those that see technology as being all-powerful and capable of solving any problem put before it, and at the other extreme those who see technology as having no place and the fragmentation and decentralisation of society as being inevitable. I started out thinking that this spectrum would be linear, but actually both extremes taken to their logical conclusion result in collapse – what David Holmgren calls the 'Atlantis scenario' – where society implodes and disintegrates.[1] As a way of underpinning this diverse spectrum of possibilities, I have used three mindsets:

- **Adaptation**: scenarios that assume we can somehow invent our way out of trouble

- **Evolution**: scenarios which require a degree of collective evolution, a change of mindset, but which assume that society, albeit in a low-energy, more localised form, will retain its coherence

- **Collapse**: scenarios that assume that the inevitable outcome of peak oil and climate change will be the fracturing and disintegration, either sudden or gradual, of society as we know it.

From the diagram (Figure 8) we can see that those scenarios in the top left hand section (the Adaptation wedge), all rely on technology, economic growth and the continuation of economic globalisation to solve the problems that peak oil is presenting. Many of them don't even allow for the mitigation of climate change. Put simply, we don't need to change ourselves, just our light bulbs. Scenario planner Pierre Wack has said that these kinds of scenarios have a fatal flaw, in that they rely on what he terms the "Three Miracles",[2] namely:

1. **A technological miracle** (i.e. extraordinary new exploration and production levels or free/hydrogen energy)

2. **A socio-political miracle** (that government policies and cultural values will allow social exclusion to be eradicated)

3. **A fiscal miracle**, namely that the public sector will fund the implementation of that scenario.

The Evolution scenarios require the actual evolution of our culture as a whole, rather than just focusing on technological solutions to 'fix' the immediate problem. This approach echoes Einstein's famous words: "We cannot solve our problems with the same thinking we used when we created them." These scenarios argue that we have to evolve our way out of this one.

It is the unlikelihood of all of Wack's three miracles occurring that leads me to believe that the Adaptation scenarios aren't going to happen, and that the Evolution ones are the most likely. Collapse is, of course, always possible, but I like to think of it as being like The Ghost of Christmas Future in Dickens' *A Christmas Carol*. That is, it shows how the future will be unless we change what we are doing. It is not inevitable. As we will explore in more detail later in this section, much of what we would need to do to prepare for the Collapse scenarios we would need to do anyway to prepare for the Evolution scenarios. I would argue that rather than trying to terrify people into change through presenting them with visions of Collapse, the Evolution scenarios could provide a vision of an end goal so enticing that society would want to engage in the transition towards them.

Scenarios in the Evolution spectrum range from the idea that what is needed is a national concerted plan of action to break away from dependence on fossil fuels (what Heinberg terms 'Powerdown') to localisation – the concept that we re-prioritise the local, and that the influence of centralised systems begins to decline. It is my assertion that when peak oil and the need to respond to climate change are factored into our responses, Business as Usual (and the other scenarios in the Adaptation paradigm) has no long-term viability.

Urban Colonies (Foresight)

A future of compact sustainable cities, with energy-efficient public transport systems, more isolated rural areas and reduced consumption.

Green Tech Stability (Holmgren)

Outlines the idea that business as usual can continue indefinitely, with renewable energy replacing conventional energy, hydrogen cars replacing existing cars, and so on.

Conventional Worlds (Gallopin)

This scenario is essentially business as usual, not deviating sharply from the present.

Business as Usual (Feasta)

Puts oil peak at 2030, with the government doing nothing to pre-empt its arrival.

Waiting for the Magic Elixir (Heinberg)

A new energy source as abundant and versatile as oil is developed, such as cold fusion or the mythological 'free energy'.

Perpetual Motion (Foresight)

A zero-emissions hi-tech hydrogen economy, assumes that globalisation is still in place, with strong demand for travel.

Techno Explosion (Holmgren)

Technology solves every problem we are currently presented with, leading to a world of holidays on the moon, unlimited nuclear cold fusion etc.

Last One Standing (Heinberg)

Describes a scenario where military force is used to secure remaining world hydrocarbon reserves, Dick Cheney's "war that will never end in our lifetimes".

Atlantis (Holmgren)

This scenario describes a sudden and catastrophic societal collapse.

Barbarisation Worlds (Gallopin)

Like Holmgren's 'Atlantis' scenario, this scenario projects a deterioration in civilisation as problems overwhelm the coping capacity of both markets and policies.

Green Tech Stability
Urban Colonies
Conventional Worlds
Business as usual
Waiting for the Magic Elixir
Perpetual Motion
Techno Explosion
Last One Standing
Atlantis
Barbarisation Worlds

ADAPTATION

COLLAPSE

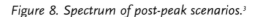

Figure 8. Spectrum of post-peak scenarios.[3]

Enlightened Transition (Feasta)

Assumes that the government decides "to use energy which is much cheaper now than it will ever be again to develop Irish energy sources and to reduce the amount of energy required to maintain and run the Irish economy". This results in an economy much more prepared for the peak when it does eventually arrive.

Powerdown (Heinberg)

"The path of co-operation, conservation and sharing", a government-led strategy utilising all the resources at its disposal to reduce per-capita consumption and build the post-fossil-fuel economy and infrastructure.

Good Intentions (Foresight)

A world in which a system of rigorous carbon rationing has been intro-duced, leading to reduced traffic volumes and more mass transportation.

Fair Shares (FEASTA)

Assumes peak oil in 2007, but a rapid government response including the introduction of carbon rationing alongside a concerted effort to reduce energy use in all areas, and the relocalisation of most aspects of daily life.

The Great Transition (Gallopin)

This scenario "incorporates visionary solutions to the sustainability challenge, including fundamental changes in the prevailing values as well as novel socio-economic arrangements".

Earth Stewardship (Holmgren)

Here "human society creatively descends the energy demand slope essentially as a 'mirror image' of the creative energy ascent that occurred between the onset of the industrial revolution and the present day".

Enforced Localisation (FEASTA)

Assumes oil peak in 2007 leading to a drastic economic downturn. The economy contracts and then collapses, resulting in a very localised future, which over time becomes increasingly sophisticated, but only within much reduced energy limitations.

Tribal Trading (Foresight)

A world that has been through a 'sharp and savage energy shock'. A global recession has left millions unemployed, and for most people, 'the world has shrunk to their own community'. Transport is typically by horse and bicycle.

Building Lifeboats (Heinberg)

Building Lifeboats "begins with the assumption that industrial civili-sation cannot be salvaged in anything like its present form" and is a process of building community solidarity, creating a localised infrastructure and preserving and enhancing the essentials of life.

EVOLUTION

COLLAPSE

- Enlightened Transition
- Powerdown
- Good Intentions
- Fair Shares
- The Great Transition
- Earth Stewardship
- Localisation
- Building Lifeboats
- Tribal Trading

(continued on facing page)

Why a future with less energy ends up looking somewhat inevitable

The work of Bryn Davidson of the Dynamic Cities Project in Vancouver BC, Canada, offers a perhaps more accessible and concise overview of post-peak scenarios. He has created a table with two axes (Figure 9). The first, from left to right, shows the rapidity with which depletion hits. As we have already seen, many indicators show that the oil peak is much closer than we think,[4] placing us almost certainly in the rapid depletion half of the table. The top to bottom axis reflects the degree of proactivity or reactivity taken by government and business.

In essence, Davidson argues that slow depletion and a reactive response would result in 'Burnout', a stubborn clinging to business as usual which results in a headlong lurch into climate chaos. If we have rapid depletion and a reactive response, the results will be cata-strophic: a societal breakdown and collapse on the scale of some of those throughout history, such as the Mayans and the Romans, as described so clearly in Jared Diamond's recent book, *Collapse*[5] or in William R. Catton's *Overshoot*.[6] Slow depletion and a proactive response could lead to what Davidson calls 'Techno Markets', that is, a technology-derived sustainable development. However, to go back to the Hirsch Report, that is only possible with at least ten years' (and preferably twenty years') notice ahead of the peak – time we almost certainly don't have. The final scenario Davidson calls 'The Lean Economy'[7] or 'Powerdown'.[8] In effect, I would argue (and Davidson's graphic puts it very clearly) that a planned and urgent energy descent is really the only desirable option left to us. Both 'Burn-out' and 'Collapse' are places we really don't want to go.

For a final way of distilling the essence of these many and varied scenarios, I turn to a

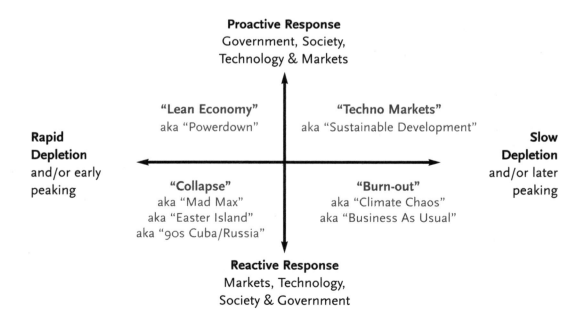

Figure 9. Four Energy Scenarios by Bryn Davidson: www.dynamiccities.squarespace.com

recent report entitled 'Descending the Oil Peak' by the City of Portland Peak Oil Task Force, which assessed the range of possible impacts of peak oil on the city, and drew from that exploration three possible scenarios.[9] They offer a concise and tidy way of looking at the realistic options facing us.

1. **Long-term Transition.** In this scenario the decline in supplies and rise in prices both occur fairly gradually, allowing mitigation options to be put in place to cope with it. It foresees a 50% cut in oil consumption over 20 years, and although the beginning of this transition is a bumpy plateau, over time the downward trend will become evident.

2. **Oil Shocks.** This scenario is similar to that above, but is punctuated by "sudden disruptions and price hikes, triggering periodic sustained emergencies".

3. **Disintegration.** Here, the impacts of peak oil become so severe that the fabric of society begins to unravel, leading to "socially catastrophic competition for scarce resources, including food, shelter and energy".

Clearly these are presented in decreasing order of desirability. The point I want to make is that preparing just for climate change and not taking peak oil into consideration offers little protection against each of the above. In each eventuality, it is the degree of resilience (a concept explored in more detail in Chapter 3) that has been in place in advance that will make the difference. Vandana Shiva, the Indian activist and advocate of sustainable agriculture, speaks of visiting some of the areas hit by the 2004 tsunami, and says that it was the villages with the greatest resilience that were up and running relatively quickly, and those that had dismantled their resilient economies in favour of an import-dependent, tourism-based model that were hit the worst:

"The indigenous tribes of Andaman and Nicobar, the Onger, the Jawaras, the Sentinelese, the Shompen, who live with a light ecological footprint, had the lowest casualties even though, in the Indian subcontinent, they were the closest to the epicentre of the earthquake."[10]

In many ways each of these three scenarios requires that we have put in place more to fall back on than we have at present. In each scenario we will need a strengthened, more localised infrastructure and the ability to have our basic needs met more locally. While Scenarios 2 and 3 are clearly less desirable, in preparing for them we actually make Scenario 1 more likely to come about. Our chances of a positive outcome are enhanced if that is what we are striving for. For many, the possibilities of Scenarios 2 and 3 inspire panic and a placing of self-preservation above more community-focused responses. This is a natural reaction but, I would venture, not a very healthy one. In difficult times we need each other more, not less.

One of the key arguments of this book is that when faced with these three scenarios, our best chance of a successful collective transition will not come from presenting people with the possibility of Scenarios 2 and 3. Psychologists Winter and Kroger write that "healthy functioning requires that we have faith that our needs will be met in the future; without this confidence, our trust in the world is damaged. Damaged trust can lead to four neurotic reactions that are likely to impact environmental behaviour: narcissism, depression, paranoia and compulsion."[11] Our best chance of dealing with climate change and peak oil will emerge from our ability to engage people in seeing the transition to Scenario 1 as an adventure, something in which they can invest their hope and

THREE MORE REASONS WHY NUCLEAR POWER WON'T GET US OUT OF THIS

4. Cost
A new programme of nuclear power would be staggeringly expensive. Amory Lovins has calculated that 10 cents invested in nuclear energy could generate 1kwh of nuclear energy, 1.2-1.7kwh windpower, 2.2-6.5kwh small cogeneration, or 10kwh of energy efficiency. Also, having sufficient money to invest so unwisely assumes an economy which is still growing, an increasingly unlikely prospect.

5. Peak Uranium
At the moment, there are about 60 years' worth of uranium left. However, if electricity generation from nuclear grows steadily, this figure will fall, to the point where if all the world's electricity were generated with nuclear, we'd have around 3 years supply left.

6. Carbon Emissions
Nuclear is often said to be a carbon-free way of generating electricity. While that may be true for the actual generation, it is not when the entire process is looked at. The mining, processing, enrichment, treatment and disposal all have significant impacts, equivalent to around one-third those of a conventional-sized gas-fired generating plant.

(For a thorough demolition of nuclear power in the context of peak oil and climate change, see Fleming, D. (2007), *The Lean Guide to Nuclear Energy: a life cycle in trouble*, The Lean Economy Connection)

"When we picture the energy climax as a spectacular but a dangerous mountain peak that we (humanity) have succeeded in climbing, the idea of descent to safety is a sensible and attractive proposition.

The climb involved heroic effort, great sacrifices, but also exhilaration and new views and possibilities at every step. There are several false peaks, but when we see the whole world laid out around us we know we are at the top. Some argue that there are higher peaks in the mists, but the weather is threatening.

The view from the top reconnects us with the wonder and majesty of the world and how it all fits together, but we cannot dally for long. We must take advantage of the view to chart our way down while we have favourable weather and daylight.

The descent will be more hazardous than the climb, and we may have to camp out on a series of plateaus to rest and sit out storms. Having been on the mountain so long, we can barely remember the home in a far-off valley that we fled as it was progressively destroyed by forces we did not understand. But we know that each step brings us closer to a sheltered valley where we can make a new home."

– David Holmgren (2002),
Permaculture: Principles and Pathways Beyond Sustainability,
Holmgren Design Services

their energy. This book will go on to explore how this could be done.

For those involved in trying to realise these 'evolutionary' scenarios, what is the nature of their role? David Korten sums it up as being "a dual role – hospice and midwife." [12] In other words, as well as helping people through the terminal decline of the current globalised oil-dependent infrastructure, we are also nurturing the emerging new localised economies that will replace it. Sharif Abdullah writes that "for the emerging society . . . our role is to compassionately assist in the birth of a new way of acting in the world. As with any birthing process, there will be some pain and trauma associated with the . . . birth. Our role is to minimise the pain and nurse the new society to full health." [13]

I believe that the only way through the monumental transition necessitated by the passing of the Age of Cheap Oil will be a rethinking of how we engage people in a transition of this scale. The tools we have had at our disposal until now are inadequate: we need a new toolkit (some of which is explored in Part Four), as well as a new way of seeing our role. One of the quotes that is most useful here comes not from an environmentalist, but an artist. The French painter and sculptor Jean Dubuffet wrote: "Art does not lie down on the bed that is made for it; it runs away as soon as one says its name; it loves to be incognito. Its best moments are when it forgets what it is called." [14]

Perhaps our work preparing communities for transition should similarly be constantly reinventing itself and forgetting what it is called: a creative, engaging, playful process, wherein we support our communities through the loss of the familiar and inspire and create a new lower energy infrastructure which is ultimately an improvement on the present.

Why 'Energy Descent'?

During this book I make frequent use of the term 'energy descent'. This may be a new term to you, so I will explain what I mean by it. When I first found out about peak oil, it puzzled me that the focus was solely on the top part of the graph, the peak itself. A raft of geologists, academics and writers were exploring the top of the classic bell curve; would it be a gentle descent, a 'bumpy plateau' or an abrupt cliff edge? No-one, however, appeared to be looking at the downward half of the graph. This struck me as being far more important than the peak itself, but it looked rather like the uncharted territory everyone was avoiding.

A society without access to fossil fuels would be able to do seventy to a hundred times less work than one with them, [15] and would, by necessity, look very different from the present. Alongside the question of when peak oil might be reached is another equally important point, that of 'net energy'. Net energy, also known as 'Energy Return on Energy Invested' (EROEI) has been defined as "the energy delivered by an energy-obtaining activity compared to the energy required to get it". [16] Oil production in the US in the 1930s had a net EROEI of over 100:1, meaning that for each unit of energy used in the extraction process, more than 100 were obtained. This is an incredible energy return, unprecedented in history. However, given the increasing amount of work we had to put into extracting oil and the increased refining needed due to the lower quality of the oil we found, by 1970 that had fallen to 30:1 and is now somewhere between 11:1 and 18:1. Globally, the average is around 20:1. [17] This is mostly for extracting oil from mature fields. The EROEI for new fields appears to be much lower.

Wind, for example, has a net energy of 11:1

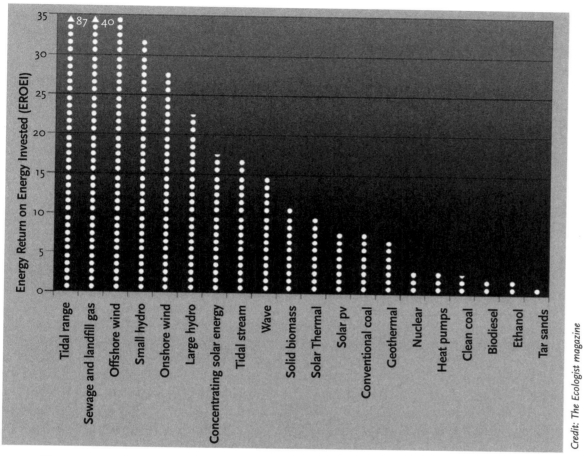

Figure 10. *Energy Return on Energy Invested for the range of energy sources. Note particularly the poor return on liquid fuels.*

"It's always going to be difficult to come up with sustainable ways to support our unsustainable lifestyle."

– Charles Wyman of the University of California on biofuels, *New Scientist*, February 2nd, 2008

(although that ratio would be much lower if corrected for the backup systems needed for when the wind is not blowing), and photovoltaics between 2.5:1 and 4.3:1. Hydro is the highest, at over 23:1, but many of the planet's potential hydro sites are already developed, many existing hydro schemes are experiencing problems from silting, and drier summers due to climate change are also reducing output in some places. For example, Costa Rica's hydroelectric capacity fell by 25% in 2007 due to the dry summer.[18] In terms of EROEI, it is the substitutes for the liq-

uid fuels that make our society possible that are revealed to offer no substitute at all. Biodiesel has a net energy of about 2:1, ethanol from sugar somewhere around 4:1 (but as high as 8:1 in Brazil, owing to their favourable climate) and corn bioethanol varies between 0.8:1 and 1.6:1. None of these comes close to oil, and Charles Hall at the State University of New York argues that to offer any remotely viable contribution to society, a liquid fuel should not be dependent on subsidies from petroleum and should have an EROEI of at least 5:1. The decline in EROEI in our energy sources, together with the combined peaks of oil, gas, coal, and uranium (probably in that order),[19] means that we need to acknowledge that we are as energy-rich a society today as we are ever likely to be.[20]

On further investigation, it turns out that a handful of people had actually begun thinking about what this journey 'down the slope' might look like. The first were Howard and Elisabeth Odum, who in their 2001 book *A Prosperous Way Down* wrote:

"That the way down can be prosperous is the exciting viewpoint whose time has come. Descent is a new frontier to approach with zeal. . . . If everyone understands the necessity of the whole society adapting to less, then society can pull together with a common mission to select what is essential. Presidents, governors, and local leaders can explain the problem and lead society in a shared mission. Millions of people the world over, if they see the opportunity, can be united in the common quest for a prosperous way down. The alternative is a world of selfish battles for whatever resources remain."[21]

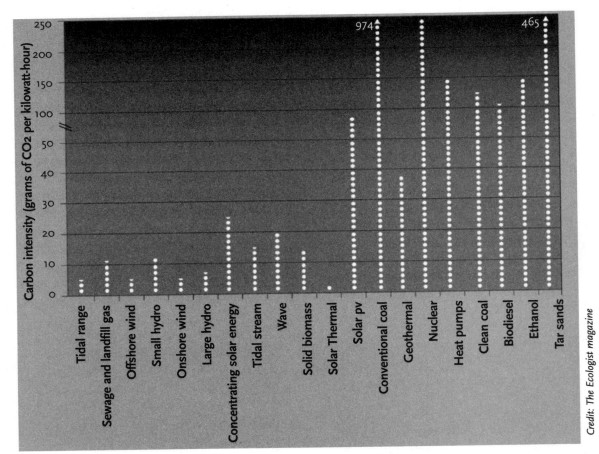

Credit: The Ecologist magazine

Figure 11. The carbon footprints for the same energy sources as grams of CO2 per kilowatt-hour. Once again this reinforces the undesirability of biofuels as well as the urgent need for coal to stay in the ground.

They argued for the need to prepare in advance for the inevitable decline in net energy availability. This echoes the Hirsch Report's stark statement that any societal-scale response to peak oil "needs to be initiated more than a decade in advance of peaking". [22]

The term 'energy descent' was further developed by David Holmgren, the co-founder of permaculture, who in 2003 wrote: "I use the term 'descent' as the least loaded word that honestly conveys the inevitable, radical reduction of material consumption and/or human numbers that will characterise the declining decades and centuries of fossil fuel abundance and availability." [23]

Most recently, Ted Trainer of the University of New South Wales has argued, in the essential 'Renewable Energy Cannot Sustain a Consumer Society', [24] that while renewable energy sources will have a key role to play beyond the peak, the idea that a Western consumer society can continue, let alone grow while being powered entirely by renewables, is absurd, and that redesigning for a far lower energy world is inevitable. He writes:

> "There is a widespread assumption that a consumer-capitalist society, based on the determination to increase production, sales, trade investment, 'living standards' and the GDP as fast as possible and indefinitely, can be run on renewable energy. . . . But if this assumption is wrong, we are in for catastrophic problems in the very near future and we should be exploring radical social alternatives urgently." [25]

In the absence of a universally agreed definition of the term 'energy descent', I would define it thus: "The continual decline in net energy supporting humanity, a decline that mirrors the ascent in net energy that has taken place since the Industrial Revolution. It also refers to a future scenario in which humanity has successfully adapted to declining net fossil fuel energy availability and has become more localised and self-reliant. It is a term favoured by people looking towards energy peak as an opportunity for positive change rather than an inevitable disaster." [26]

As Colin Campbell of the Association for the Study of Peak Oil explains, the important thing about energy descent is not when peak oil occurs, but rather the shift in perception that the Age of Easy Oil is drawing to a close. It is not so much the rate of change as the change of direction that is important. The concept of energy descent, and of the Transition approach, is a simple one: that the future with less oil could be preferable to the present, but only if sufficient creativity and imagination are applied early enough in the design of this transition.

We have a choice. We can descend the hill on which we are standing if the same imagination and drive that got us to the top in the first place can be harnessed. The reality is that the only way from here is down (in net energy terms), but that 'down' need not necessarily mean deprivation, misery and collapse. Trying to build a Heath-Robinson-style 'extension' to the hill (built on foundations of coal-to-liquids, tar sands and so on) – a rickety artificial slope that attempts to deny the geological reality of the hill itself – only means we have further to fall. The idea of energy descent is that each step back down the hill could be a step towards sanity, towards place and towards wholeness. It is a coming back to who we really are, similar to how members of a busy family rediscover each other during a power cut. Energy descent is, ultimately, about energy ascent – the re-energising of communities and culture – and is the key to our realistically embracing the possibilities of our situation rather than being overwhelmed by their challenges.

Chapter 3

Why rebuilding resilience is as important as cutting carbon emissions

What is resilience?

"A resilient system is adaptable and diverse. It has some redundancy built in. A resilient perspective acknowledges that change is constant and prediction difficult in a world that is complex and dynamic. It understands that when you manipulate the individual pieces of a system, you change that system in unintended ways. Resilience thinking is a new lens for looking at the natural world we are embedded in and the man-made world we have imposed upon it."

– Ward, C. (2007) 'Diesel-Driven Bee Slums and Impotent Turkeys: The Case for Resilience', www.tomdispatch.com

The concept of resilience is central to this book. In ecology, the term resilience refers to an ecosystem's ability to roll with external shocks and attempted enforced changes. Walker *et al.* define it thus:

> "Resilience is the capacity of a system to absorb disturbance and reorganise while undergoing change, so as to still retain essentially the same function, structure, identity and feedbacks." [1]

In the context of communities and settlements, it refers to their ability to not collapse at first sight of oil or food shortages, and to their ability to respond with adaptability to disturbance. The UK truck drivers' dispute of 2000 offers a valuable lesson here. Within the space of three days, the UK economy was brought to the brink, as it became clear that the country was about a day away from food rationing and civil unrest.

Shortly before the dispute was resolved, Sir Peter Davis, Chairman of Sainsbury's, sent a letter to Tony Blair saying that food shortages would appear in "days rather than weeks". [2] The fragility of the illusion that, as DEFRA said in a 2003 statement, "national food security is neither necessary nor is it desirable," [3] became glaringly obvious. It became clear that we no longer have any resilience left to fall back on, and are, in reality, three days away from hunger at any moment, evoking the old saying that "civilisation is only three meals deep". We have become completely reliant on the utterly unreliable, and we have no Plan B.

The concept of resilience goes far beyond the better-known concept of sustainability. A community might, for example, campaign for plastics recycling, where all of its industrial and domestic plastic waste is collected for recycling. While almost certainly being better for the environment as a whole, it adds almost no resilience to the community. Perhaps a better solution (alongside the obvious one of producing less plastic waste), would be to develop other uses for waste plastics requiring minimal processing, perhaps producing tightly compressed building blocks or an insulating product for local use. Simply collecting it and

Photo © Andy Golding

While Western Europe has relentlessly reduced its own resilience, in parts of Eastern Europe it is still a central part of people's lives. This road sign from Romania says it all.

sending it away doesn't leave the community in a stronger position, nor is it more able to respond creatively to change and shock. The same is true of some of the strategies put forward by climate change campaigns that don't take peak oil into consideration. Planting trees to create community woodlands may lock up carbon (though the science is still divided on this) and be good for biodiversity, but does little to build resilience; whereas the planting of well-designed agroforestry/food forest plantings does. The Millennium Forests initiative missed a huge opportunity to put in place a key resource: we could by now have food forests up and down the country starting to bear fruit (both metaphorically and literally).

Economist David Fleming argues that the benefits for a community with enhanced resilience will be that:

- If one part is destroyed, the shock will not ripple through the whole system

- There is wide diversity of character and solutions developed creatively in response to local circumstances

- It can meet its needs despite the substantial absence of travel and transport

- The other big infrastructures and bureaucracies of the intermediate economy are replaced by fit-for-purpose local alternatives at drastically reduced cost [4]

Increased resilience and a stronger local economy do not mean that we put a fence up around our towns and cities and refuse to allow anything in or out. It is not a rejection of commerce or somehow a return to a rose-tinted version of some imagined past. What it does mean is being more prepared for a leaner future, more self-reliant, and prioritising the local over the imported.

The three ingredients of a resilient system

According to studies of what makes ecosystems resilient,[5] there are three features that are central to a system's ability to reorganise itself following shocks. They are:

- **Diversity**
- **Modularity**
- **Tightness of Feedbacks**

Diversity relates to the number of elements that comprise a particular system, be they people, species, businesses, institutions or sources of food. The resilience of a system comes not only from the number of the species that make up that diversity, but also from the number of connections between them. Diversity also refers to the diversity of functions in our settlements (rather than just relying on one – say, tourism or mining) and a diversity of potential responses to challenges, leading to a greater flexibility. Diversity of land use – farms, market gardens, aquaculture, forest gardens, nut tree plantings, and so on – are key to the resilience of the settlement, and their erosion in recent years has paralleled the rise of monocultures, which are by definition an absence of diversity.[6]

Another meaning of diversity is that of diversity between systems. The exact set of solutions that will work in one place will not necessarily work in other places: each community will assemble its own solutions, responses and tools. This matters for two reasons. Firstly because it makes top-down approaches almost redundant, as those at the top lack the knowledge of local conditions and how to respond to them. Secondly, because resilience-building is about working on small changes to lots of niches in the place, making lots of small interventions rather than a few large ones.[7]

"Research by Thames Valley University reveals the dramatic nature of the decline of small shops in villages, market towns and district centres up and down the country. Since the 1940s, around 100,000 small shops have closed, and every year their number drops by approximately 10 per cent. Between 1995 and 2000, independent fresh food specialists – including bakers, butchers, fishmongers and greengrocers – saw their sales drop by 40% as supermarkets consolidated their grip over the food retail sector."

– New Economics Foundation, *Ghost Town Britain* report

> "In a resilient system, individual nodes – like people, companies, communities and even whole countries – are able to draw on support and resources from elsewhere, but they're also self-sufficient enough to provide for their essential needs in an emergency. Yet in our drive to hyperconnect and globalise all the world's economic and technological networks, we've forgotten the last half of this injunction."
>
> – Thomas Homer-Dixon (2007), *The Upside of Down: catastrophe, creativity and the renewal of civilisation*, Souvenir Press

> "I do not want my house to be walled in or my windows blocked. I want the cultures of all lands to be blown about the house as freely as possible. But I also refuse to be blown off my feet by any."
>
> – Mahatma Gandhi

The term **modularity**, according to ecologists Brian Walker and David Salt, relates to "the manner in which the components that make up a system are linked".[8] Towards the end of 2007, the Northern Rock bank crisis led to major problems and uncertainty in the British banking system. It was caused by over-lending to high-risk house-buyers in the US thousands of miles away, but within a short period of time one system had knocked on to another and then another, showing how the globalised networks, often trumpeted as one of globalisation's great strengths, can in fact also be one of its great weaknesses. The over-networked nature of modern, highly connected systems allow shock to travel rapidly through them, with potentially disastrous effects.

A more modular structure means that the parts of a system can more effectively self-organise in the event of shock. For example, as a result of the globalisation of the food industry, animals and animal parts are moved around

This market in Slovenia features a diversity of local produce, the fruits of a far more diverse rural agricultural economy than in more industrialised nations.

Photo © Andy Goldring

the world, leading to increased occurrences of diseases such as bird flu and foot-and-mouth disease. Reducing animal transportation and reintroducing local abattoirs and processing would lead to a more modular system, with local breeds for local markets and a much reduced risk of disease spreading with the rapidity that we have seen in recent outbreaks.

When designing energy descent pathways for Transition Initiatives, the concept of modularity is key: maximising modularity with more internal connections reduces vulnerability to any disruptions of wider networks. Local food systems, local investment models, and so on, all add to this modularity, meaning that we engage with the wider world but from an ethic of networking and information sharing rather than of mutual dependence.

Tightness of Feedbacks refers to how quickly and strongly the consequences of a change in one part of the system are felt and responded to in other parts. Walker and Salt write: "Centralised governance and globalisation can weaken feedbacks. As feedbacks lengthen, there is an increased chance of crossing a threshold without detecting it in a timely fashion."[9] In a more localised system, the results of our actions are more obvious. We don't want excessive use of pesticides or other pollutants in our area, but seem happier to be oblivious to their use in other parts of the world. In a globalised system, the feedbacks about the impacts of soil erosion, low pay and pesticide use provide weak feedback signals. Tightening feedback loops will have beneficial results, allowing us to bring the consequences of our actions closer to home, rather than their being so far from our awareness that they don't even register. When people live off the grid in terms of energy, they are more mindful about their consumption because they are closer to its generation – the feedback loop is smaller.

Life before oil wasn't all bad

These are not new ideas; rather they are the unstated principles that underpinned how things always were until the Oil Age began. It can be instructive to look back into the history of our settlements to see how people employed ingenuity and common sense before cheap fossil fuels enabled us to do without them. During the 1950s and 60s in the UK there was a concerted effort to vilify the local, the small, the simple, the rustic, the 'old-fashioned'. It is a process that happened more recently in Ireland, and is happening aggressively now in China and India. Car good, horse-drawn cart bad; concrete good, cob bad; office job good, farming bad; TV good, hearthside storytelling bad. While not wishing to romanticise the past or paint an idyllic picture of localised economies, we have come to believe either that life before oil consisted of rolling around in the mud, incest, shoving young boys up chimneys and little else; or that it was some idyllic world where everyone respected their elders and had roses over the front door.

In fact there is much that we can learn from and reclaim in our history. People were generally far more skilled and practical, local economies were more diverse and resilient, and

Harvesting apples during World War II.

people more connected to where their energy and food came from. For example, in Totnes in Devon in the 1930s, the centre of town contained a number of allotments and market gardens, which provided most of the vegetables and some of the fruit consumed in the town. Apart from the railway station, all of the businesses were owned by local people. Contrast this with a recent survey by the New Economics Foundation [10] which found that of the 103 towns and villages surveyed, 42% were what they

Harvesting with scythes in south Devon, at the turn of the last century.

called 'Clone Towns', which they defined as "one which had had the individuality of its high street shops replaced by a monochrome strip of global and national chains that mean its retail heart could easily be mistaken for dozens of other bland town centres across the county." Locally owned businesses are a dying breed, and we are only just starting to appreciate how important they are, and the resilience they give to local communities and their economies.

Of course, there was much that was miserable and debilitating, and in many ways there was a lack of personal choice that today we would find intolerable. Lives were shorter, and less comfortable. However, while not for a second advocating that we model our future on our past, I would agree that we ought not throw the baby out with the bathwater. Take a walk around the endless streets, shopping centres, car parks and tarmac expanses of present-day London, and then compare them with this section from Charles Dickens' *Great Expectations*:

> "Wemmick's house was a little wooden cottage in the midst of plots of garden, and the top of it was cut out and painted like a battery mounted with guns. . . .
>
> 'At the back, there's a pig, and there are fowls and rabbits; then, I knock together my own little frame, you see, and grow cucumbers; and you'll judge at supper what sort of a salad I can raise. So, sir,' said Wemmick, smiling again, but seriously too, as he shook his head, 'if you can suppose the little place besieged, it would hold out a devil of a time in point of provisions.'
>
> Then, he conducted me to a bower about a dozen yards off, but which was approached by such ingenious twists of path that it took quite a long time to get at; and in this retreat our glasses were already set forth. Our punch was cooling in an ornamental lake on whose margin the bower was raised. This piece of water (with an island in the middle which might have been the salad for supper) was of a circular form, and he had constructed a fountain in it, which, when you set a little mill going and took a cork out of a pipe, played to that powerful extent that it made the back of your hand quite wet.
>
> 'I am my own engineer, and my own carpenter, and my own plumber, and my own gardener, and my own Jack of all Trades,' said Wemmick, in acknowledging my compliments. 'Well; it's a good thing, you know. It brushes the Newgate cobwebs away . . .'"

Although fictional, Dickens is painting a picture of areas within walking distance of central London around 1870. Wemmick was simultaneously a consumer and a producer. Most of us have long since abandoned the latter. Nowadays

Until rail and roads took over, goods were brought into coastal towns mainly by boat. The photo above shows small cargo boats around 1900.

Photo © Totnes Image Bank and Rural Archive

we might call Mr Wemmick's set-up 'a low-impact building constructed from local materials set within a biodiverse urban edible landscape integrating protected cropping, aquaculture, chicken and pig tractoring'. By 2008 it is probably a car park.

The cake analogy

I like to use the analogy of a metaphorical cake. In Totnes, as an example, prior to the advent of the railways in the 1850s, the town and its hinterland were largely self-reliant. Its milk, cheese, meat, seasonal vegetables and fruit, as well as the bulk of its building materials and some of its fabrics were all produced locally (until the Industrial Revolution, when fabric production was moved to the north of England). What came in on small sailing boats up the River Dart to be offloaded at the Quays were Baltic timber, apples for cider from Brittany (the area drank and exported a lot of cider but didn't grow enough apples) and some wool. If, for some reason, those boats stopped coming, the area would manage. It was resilient. The cake was produced locally. and the icing and the cherries on the top were imported.

Now it is the other way round. The cake is imported from wherever in the world it can be found cheapest, and local agriculture produces the icing and the cherries on top. We have moved from the resilient to the precariously unresilient. The process of dismantling the complex and diverse rural economy that supported communities over centuries, and that was unconsciously designed on the principles of resilience has, thanks to the relentless forces of globalisation, been dismantled and thrown into the large yellow skip of history over the last 40-50 years. As ecologist Aldo Leopold observed, "Who but a fool would discard seem-ingly useless parts? To keep every cog and wheel is the first precaution of intelligent tinkering."[11] We have kept very few of the parts, and the idea that we might need some of them again is only just starting to emerge.

Echoes of a resilient past

Two examples from Totnes's recent past offer both wider insights into how our settlements functioned prior to cheap oil, and some of the strategies and infrastructure they may need to consider beyond it. In the middle of the town you'll find Heath's Nursery Car Park. While to the modern eye it looks like any other car park, what it replaced is extraordinary. It used to be a vibrant and productive urban market garden, as were another two of the town's car parks. This is not unique to Totnes; it is a pattern you'll find in any settlement. The orchards, the market gardens, the coppices, the nut trees and the fish nurseries were all grubbed out and replaced by the relentless surge of urbanisation which transformed our towns and cities. Now their legacy can only be found in the street names,

Photo © Stephen Gascoigne

So often streets are named after what was paved over to make way for them. This street in Bradford-on-Avon in Wiltshire no longer contains any coppice whatsoever.

"In the cities, apart from frightening the horses, motor vehicles competed savagely with many other road users. At the end of the last century, city streets were not just transportation arteries; they served all kinds of neighbourhood and family uses. As Clay McShane has pointed out, 'push-cart vendors brought their wares to urban housewives . . . surviving lithographs and photos show great herds of children playing in the streets, generally the only open spaces.'

City streets were often flanked with stalls selling all manner of produce. Musicians, conjurors, and other street entertainers provided the poor with cheap diversion. On to this seething and varied social scene, enter the motor car and exit many other performers! When the car first appeared on cinema screens in the 1890s it was often shown in the role of the bully, tipping over costers' barrows and fruit stalls; it may have been funny on the screen but it was not nearly so much fun on the streets."

– Julian Pettifer & Nigel Turner (1984), *Automania: Man and the Motor Car,* Collins

Tools for Transition No. 2: The Web of Resilience exercise

At the beginning of any course I teach, and also at events where I need one practical exercise to communicate the Transition concept, I use the following exercise, an adaptation of one I have used for years at the beginning of permaculture courses. I have done this with very diverse groups of people and I have never had it not work; it is always very powerful.

Divide students into groups of no more than 15 (minimum of 6, ideal around 12). If you have any more than that divide them into smaller groups. Get them to stand in as tight a circle as they can, so their shoulders are touching.

Equipment needed

One large ball of string and one sticker per person in each group (these are large parcel labels that you buy on a roll), with the names of different elements of a native woodland written as visibly as possible on these in advance. Here in the UK, my list consists of Oak Tree, Soil, Hedgerow, Badger, Worm, Dormouse, Rainfall, Owl, Leaf Litter, Fox, Robin, Wetland, Hazel, Beetle, Fungi, Blackberries, and so on. You can adapt it for the species more appropriate to your area.

Directions

The setting for this exercise is a native woodland (ideally do this exercise outdoors, in a woodland, under a large oak tree, but this is not always possible, especially in an evening class –

in this case ask people to imagine themselves in a wood).

The stickers are handed round, everyone sticks theirs to the top of their chest. The ball of string is then passed across and around the circle, the only rule being that as you pass the string to someone you must make clear what your relationship is to them.

As the string is passed around you can chip in any extra information you have on woodland ecology that is relevant about relationships between the different elements.

After a while you end up with a complex web of string between everybody. When it is finished, get everyone to pull the web tight, and then to put their hands on top of it and see how strong it is. At this stage people feel quite proud of this web they have created, and are rather pleased with themselves.

Once you have the web created you can make the following observations:

"In nature, this web of relationships is inherent in all ecosystems, and it is the diversity of relationships that makes these ecosystems work. These webs are very complex and resilient, but they are also very fragile. We intervene in them at our peril, as we can never really know what effects we are having, as we have insufficient understanding of them. While we have just done this exercise about a woodland, we could just as easily do it about a town, with the butcher, the church, the

Skilling up for Powerdown students: building a web of resilience.

schools, the farmers and so on. Before cheap oil our communities and our economies depended on these networks of relationships and connections. Cheap oil gave us the dubious 'luxury' of thinking we could live without them. People now often live with no idea who lives next-door to them. What life beyond the peak will need, and what permaculture is about, is rebuilding these connections.

"Permaculture is about re-weaving this complex web of beneficial relationships. This game is a useful tool for seeing and giving form to what we have thrown away and what cheap oil does for us."

I then walk around the circle and ask them to note how some people are holding more strings than others.

"These are the key elements of the ecosystem. When we make interventions in this system we do so at our peril. We could be a farmer who decides to clear the oak trees and drain the wetland. We could be the planners in a town with a strong local economy who decide to permit a large out-of-town supermarket. Either way we often don't see the results of this intervention immediately.

"What happens when we clear the oaks (the person who is the oaks lets go of their strings)? We can see that it doesn't make much obvious difference. So then we drain the wetland (wetland person lets theirs go). Again, it looks a bit worse but not much."

Then, using a plausible narrative ("so then the farmer did this, and then that . . ."), get people to let go of their strings one after the other; at a certain point it all collapses. The point to make is that you have no idea of knowing when that happens.

"You build the out-of-town supermarket and three years later the high street is deserted. In essence, human beings before cheap oil used good design, and networks of relationships to make things happen. Since cheap oil we have lost all that. We will need to rebuild it."

For added dramatic effect, you can brandish a pair of scissors and cut the strings! As a way of teaching people about permaculture principles and about how cheap oil has transformed our society, this can be a very powerful exercise.

'Orchard Rise', 'Nursery Lane', 'Sawpit Lane'. As James Howard Kunstler is fond of telling us,[12] places are often named after what was destroyed to make space for them.[13]

Heath's Nursery was begun by George Heath in 1920, when he bought land in the middle of the town to start a nursery. In Totnes Museum is an invoice from that year for the dismantling, moving and re-erecting of a glasshouse, the first to go up on the site. In the 1930s the business expanded, as Heath's son, also called George, came into the business, and a shop premises on the High Street was obtained.

The market garden was in two parts: the first was a large open area with one heated glasshouse where seedlings were started; the second, below it, had a series of glasshouses which were also heated. George Heath (the son) also kept pigs on the site from 1940 until the late 1950s. People I have spoken to who were at school then remember being given time off class to walk down to Heath's with the pig swill from the school kitchen, and swill was also col-lected from other local schools, shops and hos-pitals. Much of the fertility came from the local bacon factory, either in the form of pig manure or (vegetarians might choose to look away at this point) as congealed blood which was added to the water used in the glasshouses as a nutri-tious plant feed.

The nursery produced tomatoes, beetroot, cabbage, lettuce, runner beans, broad beans and also a wide range of flowers, such as chrysanthemums and dahlias, all of which were sold through the shop on the High Street. They didn't grow crops such as potatoes, as these could be grown far more competitively by local farmers, and in essence, they took up too much space. Heath's also sold seed potatoes and a large range of packeted seeds which were well patronised by local gardeners. Other ways in which the nursery was linked into the local economy included the use of wood from the local sawmill ('Reeves') to make their seed trays and the selling of produce, such as strawber-ries, from other local growers.

"It was the great difficulty of trans-porting heavy materials which led to all but the most affluent until the end of the eighteenth century to build with the materials that were most readily available near the site, even when not very durable . . . If a non-local stone were required, it was sometimes brought laboriously by wagon, but always at high cost: a documented example is furnished by the Hospital of St John at Sherborne (1438-48), for which, on five rolls of parchment, the building accounts in minute detail have fortunately sur-vived. This beautiful almshouse was mainly constructed from local oolitic limestone, but for the dressings Lias stone from Ham Hill was used. Although this entailed a journey of twelve miles, the accounts show that the cartage cost more than the stone itself."

– Alec Clifton-Taylor (1987), *The Pattern of English Building*, Faber and Faber

Not Adding Resilience	Adding Resilience
Centralised recycling	Local composting
Ornamental tree plantings (e.g. Millennium Forests)	Productive tree plantings
Sourcing organic food internationally	Local procurement specifying local production supporting emerging and new industries
Imported 'green building' materials	Specifying local building materials (cob, hemp etc.)
Low-energy buildings	The local PassivHaus (see 'A Vision for 2030', pp.115-6)
Carbon offsetting	Local community investment mechanisms
Ethical investment	Local currencies
Buying choral CDs	Singing in the local choir
Sky Sports	Playing football
Consumerism	Reciprocity

Figure 12: Adding resilience leads to a re-examination of what might be best practice.

George Heath in his Nursery, c.1970.

Heath's Nursery today.

Running a nursery like Heath's was hard work. This was a seven-days-a-week business, but it performed an invaluable service to the town and provided a reasonable living. In the early 1980s when Mr Heath retired, neither of his sons wanted to take on the business and it was gradually wound down. The glasshouses were dismantled (around the same time as they were also being dismantled in the town's other market gardens) and the land was sold to the Council, which turned it into a car park. Imagine the wonderful soil that was lost! The site is marked as productive land back as far as records exist.

In 1980 George Heath was seen as being behind the times, but he was actually about thirty years ahead of the times. His model of a localised food system that was post-carbon and based on zero food miles (indeed, it allows us to coin the term 'food feet') is one we will have to rediscover and put back in place over the next few years. But often, owing to development, this option is closed to us. Some of the site of Heath's Nursery is currently being developed, the possibility of its reinstatement gone for the foreseeable future.[14]

Going back further in local history, we find another story with insights into how society functioned before oil, as well as how it might re-organise itself beyond it. The Blight family business was horses, in particular draught-horses, which provided much of the town's horsepower (literally) prior to the arrival of the internal combustion engine. In the same way that a globalised, energy-intensive infrastructure now exists to keep motorised transport functioning, before the 1930s a localised, low-energy, diverse infrastructure existed to support the horse-powered economy. One could find a blacksmith within at most a five-mile radius of anywhere.[15] Also there were saddlers, harness-makers, ostlers, wheelwrights, grooms, ferries,

Heath's shop on the High Street, decorated for the Queen's Silver Jubilee in 1977.

Robert Blight and his son with one of his draught-horses at the rear of their home in Totnes.

> "I have always thought that the substitution of the internal combustion engine for the horse marked a very gloomy milestone in the progress of mankind."
>
> – Winston Churchill (1954)
> *Churchill Reader: a self-portrait,*
> Eyre and Spottiswoode

> "It was not only vegetables that the gardens and allotments of Britain yielded: domestic hen-keepers were supplying roughly a quarter of the country's officially known supply of fresh eggs, and there were 6,900 pig clubs throughout the country with the animals kept in straw on allotments and back gardens – and much more unlikely places including the (drained) swimming pool of the Ladies' Carlton Club in Pall Mall."
>
> – Gardiner, J. (2004)
> *Wartime Britain 1939-1945,*
> Headline Book Publishing

coachmen and vets.

The Blights' business was set up by David Blight, who started his business providing horsepower for the cutting of the South Devon railway line in the 1870s. As well as that, he also provided horses for the tramway that then ran from Totnes station to the Quays. When he died in 1889, his son Robert took over the business.

At its peak, the Blights owned eight horses, which were kept stabled in the middle of the town in what was, to all intents and purposes, the back room of their house. At that time, most of the hotels in the town had their own stables in order to be able to provide horses for carriages passing through the town. The Blights' haulage business ran until 1930, when the Totnes Fire Brigade, one of their main customers, upgraded to a motorised fire engine. At this point Robert Blight sold his interest to a local transportation company and became a manager for them.

What is interesting to me about the Blights, like George Heath, is the insight they offer into the infrastructure that was necessary before the internal combustion engine. If there was a fire, the Fire Brigade needed something to pull their engines, and they needed it urgently. Bringing the horses in from the surrounding fields would have been too time-consuming. Even though the horses were basically stabled in the Blights' house, they were a key part of the town being able to function. With the arrival of the car and the tractor, the infrastructure that supported horses contracted rapidly.

I asked a local farmer, who grew up farming in the 1930s with horses about four miles from Totnes, whether he mourned the passing of working horses. He replied: "It depended very much on the individual. If economics were your objective, then the change away from horses brought great pleasure. If you were artistic and poetic, it was a shame. I started retiring my horses in 1934 when the first tractor arrived, and I just stopped replacing them as they died out. The horsemen just became tractor drivers."

We can see the resilient rural economy, within which both of these examples sat, as being like a web of strings that connected all the various elements of the community together, similar to the 'Web of Life' exercise (see page 60). This web of connections was, while complex and resilient, very fragile. In effect, the Age of Cheap Oil took a pair of scissors to this web, replacing these functions with more energy-dependent versions. It is easy to understand why this happened and why people embraced it. We all would have done the same, had we lived in that time. It saved time, was less hard work, offered new opportunities, development, and was seen as providing a better life for the next generation. No one could have foreseen the implications fifty years down the line.

It is easy to forget the circumstances that led to many changes that we now take for granted. The move away from coal, for example, was driven as much as anything by the fact that on a bad day, city dwellers couldn't see more than a few feet in front of them, and thousands died every year from the effects of smoke inhalation.[16] However, now it is becoming clear that the cheap oil required to sustain our oil-dependent lifestyles is not going to be with us indefinitely, we find ourselves looking around at the severed strands of web and starting to wonder which strands might reconnect to which others. The Transition approach is one of re-weaving this web, and remaking the connections which will be needed by a resilient post-oil economy. Every new harmonious relationship we forge is a step back to sanity.[17]

Can we learn anything useful from Britain's last 'wartime mobilisation'?

Can any lessons be learned from Britain's most recent national 'Powerdown', World War II? While there are clearly many differences with the kind of proactive energy-descent planning this book advocates, there are also relevant similarities. We are clearly a very different society now, with different skills, expectations and values, and the nature of the challenge facing us is very different; yet even so, a look back to that time can be instructive. As Andrew Simms of the New Economics Foundation observes,

> "Recent history demonstrates that whole economies can be re-geared in short periods of time, which is exactly the demand global warming makes of us. . . . Could it be that the experience of social and military mobilisation in wartime might answer the biggest question to do with global warming: are we capable of changing our lifestyles and economies enough and in time to stop it?"[18]

There is much that can be learned from both the run-up to the war and from the 1939-45 period itself. In the light of the need for broad engagement across sectors in response to a life-threatening situation, we can learn some lessons about how quickly governments can respond (when they have to) by looking at how the British Government prepared for the impacts war would have on food production.

In April 1936, with war against Germany a possibility but by no means a certainty, an Act of Parliament set up two committees: one was commissioned to design and prepare a scheme of food rationing, and the other to propose the commodities to be given priority in a programme of storing food. This led to the creation of the Food (Defence Plans) Department in the Board of Trade, which became the driving force in preparing the food sector for war. Even so, Alan Wilt argues,[19] it was not until 1940 that the government produced a long-term policy. Committees were set up in 476 districts nationwide to co-ordinate the reorientation of agriculture. As well as attempting to increase levels of stored food, increasing home food production became a major concern. In 1936, two-thirds of Britain's food was imported and much of the nation's productive land was under pasture.[20]

Young evacuees from Acton in London making a vegetable garden at Totnes High School during World War II.

Photo © Totnes Image Bank and Rural Archive

"If I were asked the straight question whether I prefer to plough, always, with horse or tractor, I might find it difficult to answer, since I very much enjoy working with a tractor and a three furrow plough. Yet nothing can really compare with the simple, strenuous horse-work. For one thing there is no other physical work to compare with it: there is not a game in the world that can make you feel half so good. And, fascinating as the machine work is, you do not hold the plough. But it is just this grasping of the handles of the plough, both arms stretched out fully and often putting out full strength, that somehow is the very top-notch of satisfaction. Ah, I say, even as I write these lines, give me the plough-handles that I may grip them and strike out across the field! release me from this chair! (for it is so much easier to do a thing than write about it, so much easier to perform than to reveal)."

– John Stewart Collis (1973), The Worm Forgives the Plough, Penguin

Photo © Imperial War Museum

"By the end of May 1941 Bristol had more than 15,000 allotments, Nottingham 6,500 occupying 570 acres, Norwich had 4,000 over 400 acres and the number of Swansea's allotments had doubled its total since the outbreak of war, while in Tottenham in north London there were almost 3,000 allotments covering an area of 150 acres. There were allotments on school playing fields, and some factory owners encouraged employees to dig allotments on their land. The LMS railway had 22,000 which, it was estimated, would stretch from London to beyond Dumfries if put end to end, and the wife of the Keeper of Coins and Medals at the British Museum planted rows of beans, peas, onions and lettuces in the forecourt."

– Gardiner, J. (2004)
Wartime Britain 1939-1945,
Headline Book Publishing

By 1944 the amount of land under cultivation had increased from 12.9 million acres in 1939 to 19.8 million, food production had risen 91% and in effect Britain was able to feed itself for approximately 160 days a year rather than the 120 days it had been in 1939.[21] Food imports to the UK halved between 1939 and 1944. Local authorities set up horticultural committees to advise people on growing food, complemented by a huge programme of promoting the virtues of thrift and economy, as well as teaching practical skills. Some of the posters produced at the time are great examples of how to promote conservation, frugality and food production.

In 1942, Bristol (for example) had 15,000 allotments, and over half the nation's manual workers had an allotment or garden, producing around 10% of the nation's food. People sometimes remark that during the war, allotments and back gardens 'only' produced 10% of the national diet, but the important point is that the 10% it produced was the 10% that kept the nation healthy. While agriculture grew the carbohydrates and the fats, it was the back gardens that produced most of the fresh fruit and vegetables.

Food rationing was introduced on January 8th 1940, and initially applied only to bacon, butter and sugar, before being expanded to cover most foods (apart from fish and chips!) as well as fuel and clothing. One of the successes of rationing was that it rebalanced inequalities in diet. While the wealthy saw their diet restrained, for the poor, particularly in industrial centres, diet improved significantly from the pre-war years. Total food consumption fell 11% by 1944, as did meat consumption. Infant mortality rates also fell, and arguably the UK's general state of health was never better, before or since. In terms of car use, petrol rationing, introduced in 1939, was restricted to 1,800 miles per year for non-essential users, then gradually reduced until 1942 when individual allocations were abolished. Between 1938 and 1944 there was a 95% drop in the use of cars in the UK.

Much can be learned from the experience of World War II regarding how governments prepare for such a transition. The British Government was able, between 1936 when the Food (Defence Plans) Department was set up within the Board of Trade and 1939 when the war began, to co-ordinate a response which was able (just) to support the nation.[22] The most important lesson from the war years, according to Andrew Simms, is that "when governments really want to, they can do almost anything, including good things."

Clearly peak oil and climate change have yet to engender in the population or within government a sense of urgency anywhere near that of a Nazi invasion. However, as the Hirsch Report states, by the time a government considers it politically expedient to take the scale of action prompted by peak oil, it is too late. In

·· every available piece of land must be cultivated

GROW YOUR OWN FOOD
supply your own cookhouse

Photo © Imperial War Museum

"In the First World War the King, George V, had directed that potatoes, cabbages and other vegetables should replace the normal geraniums in the flower beds surrounding Buckingham Palace and in the royal parks. The Prime Minister let it be known that he was growing King Edward potatoes in his garden at Walton Heath, and the Archbishop of Canterbury issued a pastoral letter sanctioning Sunday work."

– Gardiner, J. (2004),
Wartime Britain 1939-1945,
Headline Book Publishing

terms of the model in Figure 8, the response during World War II was arguably closest to Heinberg's 'Powerdown', although the government's emphasis on local action and reskilling places it further round towards his 'Building Lifeboats'. Alongside the war effort, the building of resilience became a national priority, and was actively encouraged and facilitated by national government.

Chapter 4

Why small is inevitable

Relocalisation

A growing number of writers and thinkers now argue that the decline in availability of liquid fuels and their rising price will inevitably lead to the local scale becoming more important. As David Fleming writes, "Localisation stands, at best, at the limits of practical possibility, but it has the decisive argument in its favour that there will be no alternative."[1]

A recent report exploring the potential relocalisation of the Bay Area in California, US defines relocalisation thus:

> "The process by which a region, county, city or even neighbourhood frees itself from an overdependence on the global economy and invests its own resources to produce a significant portion of the goods, services, food and energy it consumes from its local endowment of financial, natural and human capital."[2]

I would argue that we need to be building the capability to produce locally those things that we can produce locally. It is, of course, easy to attack this idea by pointing out that some things, such as computers and frying-pans, can't be made at a local level. However, there are a lot of things we could produce locally: a wide range of seasonal fruit and vegetables, fresh fish, timber, mushrooms, dyes, many medicines, furniture, ceramics, insulation materials, soap, bread, glass, dairy products, wool and leather products, paper, building materials, perfumes and fresh flowers – to name but a few. We aren't looking to create a 'nothing in, nothing out' economy, but rather to close economic loops where possible and to produce locally what we can.

This raises enormous questions as to what a more localised manufacturing sector would look like, and the practicalities and economics of rebuilding a zero-carbon (or ideally carbon-negative) localised manufacturing sector – a sector that has been, over the past two decades, largely dismantled and outsourced to China. Although China has become a voracious consumer of oil, coal, gas and most other raw materials, more than half of the energy and raw materials it takes in are used to make products for export. When considering the reduction of the UK's carbon emissions, it is worth remembering that at the moment we don't factor in what they would look like if we started making things again, as we shall doubtless have to. The rethinking of industry on this scale and the practicalities thereof are beyond the scope of this book: I am not going to describe *how* it should happen, but rather to say that it will inevitably *need to* happen.

In our current society, everything is working against the kind of local resilience-building discussed in Chapter 3. We had a very clear example of this in Totnes when we asked the Regional Development Agency if they would fund our Local Food Directory: we were told that they couldn't, because under the rules of the World Trade Organisation they are unable to fund anything that promotes the idea that local produce is in any way superior to internationally sourced produce.

The move away from the local in the devel-

"If a significant petroleum crunch occurs, as is very likely, that will concentrate minds wonderfully. We are so extremely dependent on petroleum that any significant increase in scarcity or price will surely jolt people into the realisation that radically different social arrangements will be turned to. Without petrol it will be glaringly obvious that only localised economies will make sense."

– Ted Trainer (2007), *Renewable Energy Cannot Sustain a Consumer Society*, Springer Verlag

"Humans are only fully human when we are involved with each other, and the majority of us find happiness most easily through collective achievement. If we join our neighbours in the adventure of building a local economy that supplies and supports us all, true happiness, deep joy, is waiting to be found."

– Richard Douthwaite (1996), *Short Circuit: strengthening local economies for security in an unstable world*, Green Books

oped world was only partly by choice: some very powerful forces made it happen, and are still rampantly at work here and elsewhere. This is especially the case today in developing countries such as India, where people are being driven off the land and rural economies are systematically undermined. Ultimately, as availability of fossil fuels begins to contract so will our ability to move goods around, and inevitably, we will need to start building the infrastructure for local provision. It is not something we have a choice over – it is an emerging reality; a 'when', not an 'if'.

There will always be trade between nations, but we will be increasingly moving towards a situation where more of our core needs are locally sourced rather than imported, and the distances from which we import goods will be contracting. There is a far stronger case for importing computers and electronics than apples and chicken. Food is the most sensible place to begin rebuilding community resilience, but building materials, fabrics, timber, energy and currencies follow soon after.

The argument for decentralised living can be traced back through the communities movements of the 1960s and 70s all the way to Ebenezer Howard and William Morris in Victorian times. People such as Helena Norberg-Hodge,[3] Paul Ekins[4] and Kirkpatrick Sale[5] have more recently been arguing that the most sustainable scale of living is the local scale, and that the key to a sustainable future is the devolution of power back to communities. It has been a long-running debate, although, without external circumstances (in particular peak oil and climate change) coming to bear on the matter, it is likely that the local/global debate would end up as just that – a debate – as the forces behind globalisation are much more powerful than the forces for localisation.

Others, such as environmental activist and writer George Monbiot, have argued against the principle of blanket localisation, arguing that it is in no one's interest. The world is not equally endowed with minerals and other raw materials, so it makes sense for each place to be able to specialise. Not everywhere can manufacture saucepans, for example. Monbiot has argued that the principle of complete localisation is "coercive, destructive and unjust", and that "the money the poor world needs has to come from somewhere, and if our movement rejects trade as the answer, it is surely duty-bound to find another."[6] The point is that we are not talking about complete localisation, but rather about the building of resilience in both worlds, North and South – two processes running in parallel and in a mutually supportive way.

For Vandana Shiva, the strengthening of local economies in the developing world can only happen if agriculture relocalises in the West too. They are mutually intertwined. As she told the 2007 Soil Association conference:

> "The future of the world in farming is to produce more food in diversity, locally. And that can't be done without substituting fossil fuels for renewable energy, including human energy. Then for the first time in the last 500 years since colonialism split us into the North and South, the colonised and the coloniser, we actually have the opportunity to be one family practising a one-planet agriculture."[7]

As Vandana Shiva also points out, sustainable indigenous economies do not airfreight their produce. By the time an area is doing that, the indigenous farmers are usually long gone, cleared off the land for intensive export-driven agriculture. That's not to say that we should all close off from each other, rather that we should find more equitable and useful ways of relating

"It's easy to dismiss the principle of self-reliance by pointing to many complex products that communities cannot manufacture on their own. The goal of a self-reliant community, however, is not to create a Robinson Crusoe economy in which no resources, people or goods enter or leave. A self-reliant community simply should seek to increase control over its own economy as far as is practical."

– Michael Shuman (2000)
Going Local,
Simon & Schuster

"The political economy of the future will be lean, flexible, locally self-reliant, ingenious, robust, intelligent and very different from our own."

– David Fleming (2007),
Lean Logic: A Dictionary of Environmental Manners
(unpublished)

"The salient fact about life in the decades ahead is that it will become increasingly and intensely local and smaller in scale. It will do so steadily and by degrees as the amount of available cheap energy decreases and the global contest for it becomes more intense. The scale of all human enterprises will contract with the energy supply. We will be compelled by the circumstances of the Long Emergency to conduct the activities of daily life on a smaller scale, whether we like it or not, and the only intelligent course of action is to prepare for it."

– James Howard Kunstler (2005),
The Long Emergency; surviving
the converging catastrophes of
the twenty-first century,
Atlantic Monthly Press

in place of the unequal exchanges of 'stuff', which perpetuate rather than supersede the legacy of colonialism.

For years people have argued over the economic advantages and disadvantages of localisation. Peak oil puts an end to that debate. As David Korten puts it in his recent book *The Great Turning*:

> "People will say that 'Korten wants to change everything'. They miss the point. Everything is going to change. The question is whether we let the changes play out in increasingly destructive ways or embrace the deepening crisis as our time of opportunity. . . . It is the greatest creative challenge the species has ever faced." [8]

When peak oil is dropped into the mix, localisation is no longer a choice – it is the inevitable direction in which we are moving, one we can do nothing about, other than to decide whether we want to embrace its possibilities or cling to what we perceive that we are about to lose. The Oil Age can be seen as a 200-year period which enabled us to move away from a primarily local focus and then to move back to it again.

The principal reason for this is transportation. Peak oil is primarily a problem of liquid fuels, and liquid fuels are rarely used in the UK now to generate power. Coal tends to be used to generate electricity, gas for power and also for domestic heating, but the liquid petroleum products are key to transportation. [9] In 2004, 74% of petroleum products were used for transportation, and figures for the following year show that nearly all (98.8%) of energy consumed by the transport sector was petroleum. [10] Some of this consumption is essential, such as emergency services, public transport and agriculture, but much of it has been necessitated by work options, settlement designs, the systematic undermining of local economies over the past 50 years, and our deeply ingrained cultural perception that we have the right to go where we want, when we want, and how we want. As you can see in the quotation in the sidebar opposite, the availability of cheap liquid fuels has also allowed us to design a food supply system in which huge amounts of energy are used moving food and other goods around just for the sake of it. As Herman Daly puts it: "Exchanging recipes would surely be more efficient." [11]

Writing in the *New York Times*, Fareed Zakariah recently wrote: "You cannot switch off those forces [of economic globalisation] except at great cost to our own economic well-being. Over the last century, those countries that tried to preserve their systems, jobs, culture or traditions by keeping the rest of the world out stagnated. Those that opened themselves up to the world prospered." [12] While there is much about

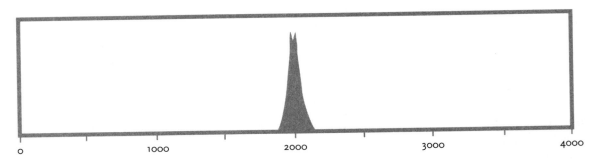

Figure 13. The Petroleum Interval in its historic context.

this statement that is contentious, its principal weakness is its reliance on cheap liquid fuels. Nothing can keep the 30 million cars and just over 2 million lorries[13] on the UK's roads indefinitely, and by extension, the more than 600 million cars in the wider world. Nothing. Complete reliance on road transport and centralised distribution are economic globalisation's Achilles' heel. Various 'alternative' fuels are proposed, generally by those for whom the 'viable mitigation option' of systematically weaning ourselves off private transport and centralised distribution does not enter the frame; but as we shall see, they fall short when examined critically.

The two most widely touted examples of alternatives to liquid fuels are:

Biodiesel

Initially, this sounds like a great idea. You grow crops, you press the oil from them, you run cars on it. In theory it is carbon neutral, and creates new livelihoods for farmers. The reality is somewhat different. To start with, there simply is not the land available to do it. To power the current national fleet would need almost 26 million hectares of arable land. The problem is that the UK only has less than six.[14]

The ethical issue that arises early in any examination of biofuels is the conflict between the use of land for food or for fuel crops. In the starkest terms, the question is whether we eat or whether we drive. With 800 million people malnourished in the world, this is a vital question. The worst example of this is corn bio-ethanol in the US, the increasing demand for which has driven the price of corn up so far that there have been riots in Mexico, with people unable to afford corn for tortillas. The land used to grow biofuels also reduces the amount of land available for growing grains for human and (far less

efficiently) animal feed. Biodiesel isn't as bad as corn bio-ethanol, but almost. In any sane society, food growing should take precedence over liquid fuels for cars every time. As David Strahan, author of *The Last Oil Shock*, succinctly puts it, "Even if we devoted all our cropland to biofuel production we would only produce a quarter of our current fuel consumption. We could all starve to death in a traffic jam."[15]

The last study that was done on whether the UK could feed itself from its land mass was in 1975.[16] It found that it was theoretically possible, but only if the diet contained a lot less meat, was similar to that of World War II, and if a lot of land not currently used were brought into production. The study concluded: "With proper planning, a little self sacrifice by the more carnivorous, and a joint effort by all sections of the community, we can build a better-fed and more beautiful Britain in the future." No mention of spare land for biofuels there.

Advocates of biodiesel come at this the wrong way round. We need to prioritise, and we need a Royal Commission on Food Security (as proposed by MEP Caroline Lucas[17, 18]) to set out those priorities. Food first, then medicinal plants and materials, then fabric crops, then building materials, and down near the bottom (just slightly above building supercasinos) biofuels, if – and it's a big if – there is any land left.

Hydrogen

Hydrogen is the science fiction fantasy of energy systems. It promises a lot: it is bold, shiny, somehow rather intrepid and of great appeal to boys of a certain age who like gadgets and toys. But like much science fiction it is a fantasy, destined to look faintly ridiculous to future generations: an illusion devoid of substance. Start looking closely, and its failings stack up rather quickly.[19] Its weakness is that it

"In 2004 the UK imported 17.2 million kilos of chocolate-covered waffles and wafers and exported 17.6 million kilos; we imported 10.2 million kilos of milk and cream by weight, from France and exported 9.9 million. The figures for the same trade with Germany were 15.5 million kilos and 17.2 million. Germany sent us 1.5 million kilos of potatoes and we sent them, yes, 1.5 million kilos of potatoes. We imported 43,000 scarves from Canada and exported 39,000. Drink is swilling around the international markets. The UK imported £310 million worth of beer in 2004 and exported £313 million worth. For spirits the figures were £344 million and £463 million respectively. Just as we imported 44,000 tonnes of frozen boneless cuts of chicken, we exported 51,000 tonnes of fresh boneless chicken."

– Andrew Simms et al (2006), *The UK Interdependence Report*, New Economics Foundation

Photo © www.istockphoto.com

The car of the future – or an over-hyped and unfeasible fantasy? Hydrogen offers the illusion that we can keep motoring into the distant future, but where will the electricity required to create the hydrogen come from?

is not an energy source, but an energy carrier. In order to obtain pure hydrogen, we need to run electricity through water. Hydrogen is not something one finds lying around in pools just below the Earth's surface from which it can be conveniently harvested and poured into our fuel tanks. The problem arises in how we generate that electricity. David Strahan has estimated that just running the UK's cars on hydrogen would necessitate "67 Sizewell B nuclear power stations, a solar array covering every inch of Norfolk and Derbyshire combined, or a wind farm bigger than the entire southwest region of England"[20] – none of which is desirable nor feasible, and any renewable energy thus generated would be more wisely used in the residential sector.

In oral histories I have done locally around Totnes, one of the recurring themes is how during World War II, when petrol rationing came in, people became far more reliant on those around them: the farmers, the craftspeople, the next-door neighbours. As the price of liquid fuels begins to rise here, I believe that we will see the focus returning once more to the local. As the forces that have undermined and ravaged local economies begin to wane, we will see a resurgence of the human-scale, of the appropriately scaled. That's not to say that some importing and exporting won't happen; it always has and it always will. This is a good time to be investing in commercial-scale wind-powered sailing ships (indeed, some people already are). As the price of energy rises, it will be interesting to see which things first become viable to produce locally again. The solutions that emerge will be dependent on our collective efforts, and on other local circumstances such as the availability of arable land and other resources.

It is clear that throughout history it made more sense to produce what was possible locally, and to import luxury goods and the few things we were unable to produce ourselves. Indeed, to do otherwise was utterly impractical and financially out of the reach of most people. The key issue here, once again, is resilience. With that resilience in place, if computers and plastic toilet brushes stop coming in, we'll still have sufficient food, shelter, fuel, basic goods and medicines to get by. Things would not be ideal from our current-day perspective, but neither would they be catastrophic. As in most towns and villages prior to 1850, imports would be things that improve our quality of life and provide products and materials we are unable to produce here, but without which we would manage without the risk of destitution and starvation.

To recap, given that our current globalised/centralised supply systems are entirely dependent on cheap liquid fossil fuels, and the uninterrupted supply of those fuels and their continuing cheapness is increasingly in doubt, we need to refocus on the creation of local production systems. Unfortunately they have been systematically and relentlessly vilified and undermined over the last sixty years. As James Howard Kunstler has written, the future will be "increasingly and intensely local and smaller in scale".[21] However, I do not wish to be seen as idealising local communities: I have lived in deeply insular rural communities, and am familiar with their not-so-good as well as their good qualities.

In the same way that the privations of World War II led to a renaissance of the fortunes of British farming (still talked of as a 'Golden Age'), peak oil and gas could lead to a renaissance of agriculture and of local low-impact manufacturing in the UK. We cannot go back, nor would we want to. We need not all relearn Morris dancing, deprive women of the vote, or re-embrace feudalism. We can adapt our culture to a more local context with creativity, and the results will be beyond our current imaginings.

What is inevitable, though, is the return of the local and the small-scale, and the turning away from the globalised. This will not be an isolationist process of turning our backs on the global community. Rather it will be one of communities and nations meeting each other not from a place of mutual dependency, but of increased resilience.

The dangers of clinging to the illusion of large-scale

Moving towards the kind of low-energy, more localised future outlined in this book is not, of course, the only option on the table. Indeed it is, at present, a table groaning under the weight of various impractical and potentially nightmarish solutions. It is worth pausing here to consider the powerful trends and forces at work taking us in an altogether different direction. There are two Transitions at work in the world: that which is the subject of this book, and another far larger, more powerful and better resourced one, which is rapidly dismantling what resilience remains, under the guise of economic globalisation and growth. Although the world's future supplies of cheap oil and gas are beginning to look very vulnerable, there is probably enough coal to make the term 'end of the fossil fuel era' somewhat redundant.

There is great debate over the oft-cited 'fact' that the world still has many hundreds of years of coal left. Many nations that previously declared huge reserves of coal are now downsizing their reserves hugely.[22] However, the likelihood is that there is still enough to irreversibly damage the climate and, via the process of converting coal to liquids, to keep the developed nations which can afford it motoring for some time yet. Indeed, the evidence that the UK Government has decided that a good part of the UK's future energy security lies in a revival of the coal industry appears to be mounting.[23] As Jeremy Leggett puts it, deciding to use the world's remaining coal reserves is to take up the challenge of trying to prove the world's climate scientists wrong.[24] It is not a contest we have much chance of winning.

If everything is left to a market unrestrained by stringent international climate change legislation, supplies of liquid fuels will be extracted from wherever they can be squeezed in a futile effort to meet demand *à la* Hirsch. We are already seeing clearance of rainforest and other ecosystems in South Asia to plant palm trees for biodiesel, to be exported to the West as a 'green' fuel.[25] We are seeing a rapidly accelerating installation of coal-to-liquids technology (making petrol from coal) around the world. We are seeing the gap open-

"A race that neglects or despises this primitive gift, that fears the touch of the soil, that has no footpaths, no community of ownership in the land which they enjoy, that warns off the walker as a trespasser, that knows no way but the highway, the carriageway, that forgets the stile, the footbridge . . . is a fair way to far more serious degeneracy."

– John Burroughs (1875), 'The Exhilarations of the Road in Winter Sunshine', from *In Praise of Walking*, Houghton Osgood

HOW LONG MIGHT KING COAL REIGN?

Coal is the monster hiding behind the bushes on the downward side of the peak oil mountain. Most people argue that there . is still 300 years'-worth of coal left, enough to boil us alive in our own climate. However, a recent report questions this assumption. The Energy Watch Group's report, 'Coal: Resources and Future Production', revisits national coal reserve data around the world and concludes that "the data quality is very unreliable". Many nations have not updated their reserves for some time, in some cases as far back as the 1960s. Excepting Australia and India, since 1986 all major coal-producing nations who have gone back and reassessed their reserves have reported what the report calls "substantial downward resource revisions".

Some, such as the UK, Botswana and Germany, have downgraded their reserves by over 90%, and Poland by 50%. In total, the world has, over the last 25 years, revised its total reserves downward by around 60%. This is likely due to improved survey techniques enabling better data to be collected. This, combined with the voracious and rapidly growing appetite for coal in the US, China and India, may well mean that 'peak coal' is much closer than we may previously have thought.

– Energy Watch Group (2007), 'Coal: Resources and Future Production', downloadable from www.energywatchgroup.org/ files/Coalreport.pdf

ing up in the UK's gas supplies being filled by an increased consumption of coal. In the winter of 2005-6, when the UK faced a substantial gas shortage, it only got through the winter by burning 18% more coal than it would otherwise have done.[26] We can see the growing demand for liquid fuels in China leading to their making oil access deals with governments and regimes that our foreign policy wouldn't allow us to. Keeping our petrol tanks full is a dirty business.

If we refuse to acknowledge the constraints that peak oil is imposing on us, we will desperately attempt to keep the economic growth show on the road for as long as possible. We will see, indeed we are already seeing, biofuels production on a staggering scale, requiring large biodiesel refineries.[27] We will also see, in parallel, rates of malnutrition and starvation rise sharply as more and more land is taken out of food production. We are seeing a vicious example of this now in Mexico. The Cantarell oil field, responsible for 60% of Mexican oil production, has begun the plunge into terminal decline. The revenue for 40% of Mexico's public funding comes from the sale of that oil to the United States. Cantarell's depletion rates are staggering. It fell around 60% from its peak between 2006 and 2007, and is estimated to fall 75% by the end of 2008.

At the same time, the US uses 20% of its corn harvest to make corn ethanol in an attempt to boost energy security. As its oil imports from Mexico continue to fall, the pressure will be for the US to produce more ethanol to fill the emerging gap, leading to falling and increasingly expensive corn imports for Mexico.[28]

Perhaps carbon capture and storage (CCS), which, in theory, captures the CO_2 from coal burning and stores it under the sea, will work, and coal will be our saviour. However, CCS is still at the experimental stage, and even if it does work, there are doubts around it being rolled out in time, about there being enough coal to make it worthwhile, and about its cost.

Our failure to adequately address the issue of climate change will have had clear consequences, with significant changes to the UK coastline and increasing extreme weather episodes, and severe ecological breakdown, mass migration and economic disruption elsewhere. Increased military spending will be required in order to sustain the dwindling supplies of conventional oil and gas. Dick Cheney's "war that will never end in our lifetimes" will have come to pass. It is a nightmare scenario, to be avoided at all costs.

Where we might be heading in terms of large-scale agriculture can be glimpsed in an article in the *Sunday Times* by Richard Girling.[29] He argues that the future of agriculture, in the light of population growth, global warming and the energy gap will necessitate the end of farming as we know it. We can say goodbye to cows standing in fields chewing the cud. We shall inhabit, he predicts, a "drive-thru, wipe-clean, prairie Britain". The countryside will be filled with "identical suburbs of identical brick-box housing", "the drone of traffic", and "vast fields of identical crops". The pressure to feed more people on less land will necessitate putting productivity above all other considerations. "There are only two ways to get more food from the soil: by breeding heavier-yielding crops or by cultivating more land," Girling argues, although, as we shall see in Part Two, there are more than just the two ways he identifies. He quotes Mark Hill of Deloitte, who says: "The challenge to farmers is to double food production in the next 40 years. How can they do it?" How indeed.

Girling argues that we will, as in 1939, need to plough up all the land where it is feasible to do so, massively reduce the amount of livestock, and embrace genetic crops as the only way through these converging challenges. This is really a point at which Einstein's saying that "problems cannot be solved by the same level of thinking that created them" becomes useful. Trying to map agriculture's journey through this transition will not be possible without some fresh thinking and an approach which actually involves asking the right questions.

Yes, of course more land will need to be put into production. The ridiculous concept of land being 'set-aside', now thankfully being phased out,[30] was a strange side-effect of a country with too much oil and supermarket shelves groaning with a dazzling array of cheap imported goods to choose from. Reducing the amount of livestock will also be inevitable, as large-scale meat production is an absurd and unsustainable waste of resources. When thinking of what a post-peak agriculture will look like, it is important to question whether, as the Once-ler in *The Lorax* might have put it, "biggering and biggering" what we have done thus far is really the best idea. The bottom line is that Girling's vision, however inevitable he might paint it as being, offers no resilience, places all our eggs in one basket, leaves us at the mercy of international events and economics, perpetuates our collective de-skilling, offers meaningful work to no one, maintains and possibly even increases the oil dependency of agriculture, destroys biodiversity, does nothing to strengthen economies, fails to make us any healthier, and would be soul-destroying. Yes, there may be some trends towards large-scale agriculture, but they are not, by any means, inevitable.

Top-down or bottom-up?

All of this reinforces why it is so important that we weave peak oil and climate change together in our decision making, and see them as being intrinsically linked. They are not separate issues, and as we saw in Figure 7, it is only by considering them together that our solutions will have any hope of being effective. Transition Initiatives will function best in the context of a combination of top-down and bottom-up responses, none of which can address the chal-

"Basically all the geological storage options have a risk of leakage; either though unsealed or insufficiently sealed drill holes or galleries (oil and gas fields and coal mines), along unknown or newly appearing faults or interruptions in channels in the geological formation intended for storage or though leakages caused by seismic activity. These can lead to CO_2 coming to the surface or into other rock strata such as those through which groundwater flows. In aquifers the introduction of CO_2 can lead to the acidification of the water already present and through this can corrode rock formations and the seals on tunnels holes that were not closed up to allow for this type of corrosion. CO_2 storage in deep coal seams carries the risk of displacing methane which has a much higher greenhouse gas potential than CO_2."

– Viebahn, P. et al (2007), 'Comparison of carbon capture and storage with renewable energy technologies regarding structural, economic, and ecological aspects in Germany', Int. J. Greenhouse Gas Control 1

International
Strong international climate change protocols, Contraction and Convergence, a moratorium on biodiesel production, Oil Depletion Protocol, rethinking economic growth, biodiversity protection
National
Strong climate change legislation, Tradable Energy Quotas, a National Food Security Strategy, Devolution of Powers to Local Communities
Local
Transition Initiatives, Energy Descent Plans, Climate-Friendly Communities, Community-Supported Agriculture, Land Trusts, Credit Unions, Locally-owned Energy Supply Companies (ESCOs), localism

Figure 14: Top-down and bottom-up responses: we'll need them all.

lenge in isolation.

The reality is that many of these responses are on their way, and are moving faster than we would have thought even a year ago. Serious thought appears to be being given at governmental level to the introduction of carbon rationing. After all, as David Fleming (the 'inventor' of Tradable Energy Quotas [31]) observes, as oil and gas production start to deplete, their rationing will be inevitable: either an equitable rationing system will be introduced, or energy will be rationed by price, which is socially divisive. It is not a question of *if*, but *when* rationing begins, and the sooner we do it, the gentler it will be. There is increasing pressure and international diplomacy around the need for strong international action on climate change, way beyond the woefully inadequate Kyoto Protocol. On an individual level, we should offer our support to any campaigns that drive forward any of the above, and direct our spending power when we go shopping to support businesses with a genuine commitment to lower energy use and sustainable business practice, in particular those whose practices build local resilience.

However, the important point is that we don't need to wait for the above. Indeed, successful national and international responses are all more likely in an environment where community responses are abundant and vibrant. We can't wait for governments to take the lead here. The UK Government's position on peak oil (i.e. complete denial) is proof of this. The Department of Environment, Food and Rural Affairs (DEFRA), put it succinctly in 2002 when they wrote: "Sustainable development cannot be imposed from above. It will not take root unless people across the country are actively engaged." [32]

Where does government fit in?

It appears to me that there is a fracture in politics. The UK Government looks to the public and sees them as disengaged, apathetic and uninterested in the democratic process. The public often sees politicians as uncaring careerists who don't have any interest in them or what is actually happening in their communities, apart from once every four years when an election comes round. Local planning consultation processes generate a lukewarm response at best. All this is happening at a time when, as we have seen, we need to be generating a response on a previously unseen scale: mobilising individuals, communities, businesses, organisations and government institutions to work as effectively together as possible in order to maximise the chances of a smooth transition.

Governments generally don't lead, they respond. They are reactive, not proactive. It is essential that we remember that many of the decisions they will inevitably have to make as part of preparing for Powerdown are perceived to be pretty much inconceivable from an electoral perspective. Take carbon rationing, for instance: few people would be brave enough at this stage to run for government on a ticket which promises people less every year – less car use and less energy availability. However, I see no reason why these ideas could not be made attractive to the electorate by the right candidates. If, through the creation of an Energy Descent Plan which has engaged the community and which offers a positive vision of a lower-energy future, communities have set out where they want to go, then a very dynamic interface is created between communities, local and national government. Communities could set the agenda, saying to government, "Here is our plan: it addresses all of the issues raised by the coming challenges of climate change and energy security, and it also will revitalise our local economy and our agricultural hinterland, but it will work far better if carbon rationing is in place, and if the true costs of fossil fuels are reflected in goods and services." The fear of change is removed for government, and they become swept along in a huge movement for change. Previously non-vote-winning policies become the norm.

Recently, corporations have begun saying to the British Government, "We want you to start taking strong action on climate change, because we need to be able to start planning for this, and we need to know the framework within which we are operating." Communities should be doing this too. We have to remember that we can do a huge amount without government, but we can also do a great deal more with them.

Summing up Part One:
the oil age draws to a close

We have seen in this first part of the book that the converging challenges of peak oil and climate change mean that change – on an almost unimaginable scale – is coming whether we want it or not. There is no longer a cosy 'if' to wrap around ourselves. We cannot adequately address the challenge of de-carbonising our society without also addressing the need to rebuild local resilience, to create local economies capable of supporting us in a post-peak world. Just cutting carbon emissions, although deeply urgent, is not enough on its own. I have delivered this message many times, in talks, courses and blog posts, and have yet to encounter anyone who thinks that stronger local economies, increased local democracy, strengthened local food culture and more local energy provision are a bad idea. Most of us instinctively know that we are living beyond our collective means, and have some sense of what we need to do.

The key message here has been that the future with less oil could be better than the present, but only if we engage in designing this transition with sufficient creativity and imagination. We need to draw together a diversity of individuals and organisations that has seldom been managed in the past. We need to employ that same adaptability, creativity and ingenuity that got us up to the top of the peak in the first place to design a way down the other side. The question now is how can we overcome the obstacles to this Transition that we encounter, both within ourselves and in the wider world?

When you hear the words climate change and peak oil, how do they affect you? What urges or instincts do they provoke? At a guess, if you are like most of the people I have asked, they lead to feelings of disempowerment, sadness, weariness, and of being confronted by something huge and scary that you feel unable to influence. This state of mind is not the place to start from, if we want to achieve something, do something or create something. It doesn't tend to lead to dynamic applied action, but we need to feel motivated and inspired. In short, we stand on the edge of the most momentous task in history, thoroughly ill-prepared.

"It was only a short time ago, two centuries at most, that we fell into our energy addiction and started down the path to ruin. Peak oil is an opportunity to pause, to think through our present course, and to adjust to a saner path for the future. We had best face facts: we really have no choice. Peak oil is a horrible predicament. It is also a wonderful opportunity to do a lot better. Let's not squander this moment."

*– Albert Bates (2006),
The Post-Petroleum Survival
Guide and Cookbook,
New Society Publishers*

78

Part Two

THE HEART
Why having a positive vision is crucial

"It is best to think of this as a revolution, not of guns, but of consciousness, which will be won by seizing the key myths, archetypes, eschatologies and ecstasies so that life won't seem worth living unless one is on the transforming energy's side." – Gary Snyder [1]

"To save the planet, we do not need miraculous technical breakthroughs, or vast amounts of capital. Essentially we need a radical change in our thinking and behaviour." – Ted Trainer [2]

"The uncertainty of our times is no reason to be certain about hopelessness." – Vandana Shiva [3]

Peak oil and climate change can be intense and distressing, both in their implications and in the effect they will have on us. In the same way that most people remember where they were on September 11th 2001, or (for the older readers) when Kennedy was assassinated, most people who are aware of peak oil and climate change have their stories to tell about the moment when 'the penny dropped' – what I sometimes call their '*End of Suburbia* moment'. I think that alongside an understanding of the issues, it is important not to pretend that we can keep our awareness of these issues on a purely intellectual 'head' level, but that we need to address the 'heart' too, acknowledging that this is disturbing information, that it affects us, and that how it affects us in turn shapes how we respond – or don't.

Also important (and explored in this section) is the concept of visioning, and the power that a vision of the future can have. Too often environmentalists try to engage people in action by painting apocalyptic visions of the future as a way of scaring them into action. The question this part of the book asks is what would happen if we came at this the other way round, painting a picture of the future so enticing that people instinctively feel drawn towards it.

As my contribution towards this, I will set out a vision for how the UK might be in 2030 if we engage creatively with this process of adaptation to energy descent, seeing our future in increased resilience, more localised economies and greatly reduced energy consumption. Inherent within the twin challenges of peak oil and climate change is an extraordinary opportunity to reinvent, rethink and rebuild the world around us. Ultimately, at the heart of this section is the understanding that the scale of this transition requires particular inner resources, not just an abstract intellectual understanding. This is relatively new ground for the environmental movement, but it is crucial to our success and to engage enough people on the scale required.

Facing page: Urban food production as an integral part of the urban fabric in Slovenia.
Photo © Andy Goldring

How peak oil and climate change affect us
'Post-petroleum stress disorder'

Before we go any further, it is worth pausing to reflect on how all of this thinking about peak oil and the changes ahead is affecting us. Having been around the subject of peak oil for a while, I have observed many people go through the process of becoming aware of peak oil, having what I sometimes call their '*End of Suburbia* moment', and have seen how that awareness affects them. For some it is a traumatic shock, for others an affirmation of what they have always suspected. For many though, it is not so clear-cut either way. I have noticed, over the years, certain symptoms of what I have come to call 'post-petroleum stress disorder'. Perhaps you might recognise some of them:

Clammy palms or nausea and mild palpitations

Finding out about something with such profound ramifications for the way we live can be a profound shock to the system. There are certain ways our bodies respond to this, and for many people the first manifestation of this disorder is physical discomfort.

A sense of bewilderment and unreality

Many spiritual traditions speak of a 'dark night of the soul', when the nature of the emptiness of reality is revealed and we are forced to let go of the understanding to which we have become attached. Peak oil and climate change put a mirror up to our lives and the society around us, enabling us to see that what we had seen as being permanent and real is in fact a fragile illusion, dependent on long supply lines and an uninterrupted flow of cheap oil. When you see the illusory nature of the world around you, it can leave you feeling bewildered.

I remember a science fiction film I saw years ago called *They Live*, which started with a man finding a box of sunglasses behind some dustbins. When he put a pair on, he could see that many of the people around him were in fact aliens who were in the process of taking over the Earth. Whenever he looked at advertising billboards, what had read 'Drink X, it'll make you happy', now read 'Consume and Die' and so on. Unfortunately, the film then degenerated into a relentless blasting of aliens with guns, but it was a very powerful metaphor for what an emerging awareness of peak oil does to our perception of the world.

An irrational grasping at unfeasible solutions

"Aha!" some people say, "it'll all be fine because we'll just switch to hydrogen!" Or nuclear power, or free energy machines made using technology recovered from UFOs. We are even told that hundreds of earnest souls, beavering away in their garages, have created devices that can yield untold quantities of free energy, in complete disregard of the laws of thermodynamics, but that every time they are about to launch them publicly, they are bought out by oil companies and the plans are put in a drawer; or that other, more sinister things happen to them. Either that, one might say, or actually they just never invented them. Or they didn't work.

Anyway, for those suffering from this symptom, there is a confident belief that there is a silver bullet out there that will enable business-as-usual to continue uninterrupted, steadily growing our economies *ad infinitum*. As we saw in Part One, the more one looks at it, there is no single technology that can enable us to continue as we are. Nuclear, hydrogen, 'clean' coal and biofuels all have severe limitations. Fossil fuels have been a one-off energy bonanza that nothing else can replace (indeed, it has been argued that we are close to 'Peak Everything', to use the name of Richard Heinberg's latest book[4]). That doesn't prevent people grasping at things that simply won't work. It is what Heinberg calls 'Waiting for the Magic Elixir'.[5] The reality is that all the technologies and appliances that we will need are already out there in the world; we just have to get on with it, rather than fantasising about impending wonder technologies.

Fear

In our work promoting responses to peak oil and climate change, we should not lose sight of the fact that for many people this is a very frightening subject. Indeed, one could argue that if you don't find it scary, you haven't really got it. For some, that fear can be paralysing, and for others it can trigger a shut-off mechanism.

It is important that we don't just dump potentially scary information on people, but rather we need to allow an exchange of information and room for people to digest what they have been told.

Outbreaks of nihilism and/or survivalism

For some, peak oil can affirm their long-held belief that people are inherently selfish anyway and what is the point – we've all had it. The survivalist response differs in that rather than thinking it is not worth doing anything, it assumes that one should prioritise self and loved ones above all else, that one should design for one's own survival; that a 'head for the hills' response is a valid one. This response is a particularly North American one, as I found out in response to a piece I wrote called 'Why the survivalists have got it wrong'. It elicited more comments than any other previous piece on Transition culture, offering a fascinating insight into those for whom individual survivalism is seen as a viable option. Some came via survivalist websites which featured such gems as "Which is better, a gun or a club? You can use a gun as a club, but you can't use a club as a gun." Of course in the US, heading for the hills is more of an option, in the UK we simply don't have the space, and decanting *en masse* to Dartmoor or Snowdonia would be a fairly unrewarding process. The most usual manifestation of this symptom that one encounters in the UK is "Well, we'll be OK, we've got a little place in the Pyrenees."

Ultimately, any response that is sufficient to the scale of the challenge is about coming home, about being aware that we are a part of the networks around us, and that we need to nurture and rebuild them, rather than imagining that we can survive independently of them. Indeed, we could see a belief that we can exist and flourish independently of the communities around us as being a dubious 'luxury' of the Age of Cheap Oil. We will have to learn to meet and greet each other once again, as well as learning how to co-operate and communicate.

Denial

In this time when climate change and peak oil are so rapidly entering the public consciousness, and the implications of what they will mean are starting to sink in, inevitably, for some people, denial comes to the fore. This can take many forms. It could be the man I sat next to on a bus who told me he had seen a TV programme where a scientist had said the world was, in fact, warming from the inside out (!), or it could be those who say that climate change is caused by sunspots or by natural cycles, despite the wealth of evidence to the contrary. It could be those who say that climate change is actually a conspiracy cooked up by the New World Order in order to further curb our freedoms, or that peak oil is a conspiracy by the oil industry to allow them to make more money. The internet is full of half-truths and Chinese whispers for those who wish to construct such denial mechanisms.

One of my favourite denial stories comes from my friend Graham Strouts,[6] who tells of a conversation with a woman with whom he had been discussing the impacts peak oil will have on the food supply system. They discussed how oil-dependent food is, and how vulnerable the system is. Then, to Graham's amazement she said, "Well it doesn't worry me – my husband didn't eat for a year once." Didn't eat for a year?! Apparently he had done some kind of meditation practice and she was convinced he hadn't

eaten for a year. Of course the assertion that it is a fairly well established scientific fact that if you take the food away from a population, they tend to start keeling over after a few weeks didn't do too much to change her position.

There is no way of completely avoiding denial, as none of us is beyond it. It pops up in all kinds of unexpected guises, and it is a natural reaction; we can't go around thinking about climate change, peak oil and the end of economic globalisation ALL the time after all! It becomes a problem when it closes us to the realities of the issue, and inhibits our ability to respond. Denial is a natural response, but we need to remain vigilant to it.

Exuberant optimism

At the end of the first public screening of *The End of Suburbia* that I organised in Ireland, a man in the audience said: "We've just been told the Oil Age is coming to a close, to which I say, 'bring it on!'." While I can appreciate the sentiment he expressed, it is not quite so straightforward. As the Hirsch Report identifies, to make the transition away from the oil-based economy will require at least ten years, preferably twenty, and a failure to adequately prepare would be disastrous. Responding to peak oil with exuberant optimism needs to be balanced with an appreciation of the massive challenge it presents.

The 'I always told you so' syndrome

I must confess that I see this one in myself. Having been involved in permaculture and natural building for many years, I naturally see peak oil as the opportunity to roll out permaculture and straw/clay houses with hemp plasters on a previously unimagined scale. For those interested in organics, peak oil is seen as the opportunity to really step up organics on a far bigger scale. The home-schoolers, the off-the-gridders, the market gardeners and the home composters may well all have their "I told you so" moments. This, for me, is entirely laudable, and a natural reaction for those who have for years been promoting various aspects of the post-carbon society years ahead of time. However, there are other people out there who are waiting to use peak oil for their 'I Always Told You So' moments from not such benign motivations.

The British National Party have taken to the issue of peak oil with great gusto, popping up at peak oil gatherings and declaring that within the peak oil challenge are the seeds to their ascendancy. Historically, fascists have always preyed on times of economic collapse and hardship, and this one is no different. One needs to take any claims of "I told you so" (including mine) with a questioning mind and great discernment.

It is clear, for example, that for the makers of the classic peak oil film *The End of Suburbia*, who were no fans of suburbia to start with, peak oil provided a great opportunity to dance upon its grave. The temptation to say "I told you so" can also mean that we neglect to really analyse the strengths and weaknesses of our proposed solutions in the context of diminishing net energy. We need to really think through the implications – in a low-energy context – of our proposals, and not remain too attached to our long-cherished beliefs and ideas. We may find instead that by letting go of them we actually come up with something better and more appropriate to a culture in transition.

COPING WITH YOUR 'END OF SUBURBIA MOMENT'

How might one best manage the feelings of overwhelm, devastation and defeat that can accompany your 'End of Suburbia moment', the point when you really 'get' peak oil and its implications? The first point is to realise that feeling like this is natural, indeed it is far more natural than feeling nothing or blanking it out. It is a healthy response.

Secondly, seek to generate what Chris Johnstone calls 'Inspirational Dissatisfaction', where the feelings generated motivate you to make changes in your life. Acknowledge that the change you want to see starts with you, and see the feeling that your life has been turned on its head as a precious opportunity to rethink some basic assumptions.

Finally, don't rush it, take some time to just sit with this new awareness. Although it feels uncomfortable, within it lies, as in the bleak opening chapters of most adventure stories, a call to adventure. You will come to look back on this as a major, but positive, transition in your life.

Chapter 6

Understanding the Psychology of Change

"Creating the world we want is a much more subtle but more powerful mode of operation than destroying the one we don't want."

– Marianne Williamson

Enabling change has always been the Holy Grail of environmentalists, but it has largely remained frustratingly elusive. Although there have been successes, overall the environmental movement has failed to engage people on a large scale in the process of change, certainly not on the scale of the wartime mobilisation now necessitated by peak oil and climate change. It could be argued that one of the reasons for this is that we have never really understood change, how it happens and what it entails. There are other disciplines that have a much better understanding of change, how it works and how to bring it about. One of these is the field of addiction.

A year ago I came across a book which transformed my thinking on this. Indeed, if it had had a different cover, and had been called something like *Peak Oil and Change for Communities*, I would have thought it a work of genius. It was *Addiction and Change* [1] by Carlo DiClemente. DiClemente developed what he calls the 'Transtheoretical Change Model' (TTM), which sets out to explain how change happens. He states that the process by which an individual gets into and out of addiction is the same as any process of change. The TTM emerges from a synthesis of various previous approaches, as well as from longitudinal studies into how people change. Rather than change being just a process of deciding to change and then changing, DiClemente argues that the process is more subtle and sophisticated than that. I found his

insights enormously illuminating, and around the same time I met Dr Chris Johnstone, an addictions specialist who has done a lot of work with the Stages of Change and applying them to social and environmental change work.

Chris is author of *Find your Power*, [2] and also edits *The Great Turning Times*, [3] as well as running workshops on 'The Work That Reconnects' around the UK. I find his take on change very inspiring and insightful, and the Transition approach is, in many ways, informed by some of these insights. Rather than trying to explain this to you myself, when I have no background in psychology or addictions, I have delegated this to Chris, and what follows is a dialogue we had exploring these issues.

An interview with Dr Chris Johnstone

What are the Stages of Change, and where did they come from?

The Stages of Change model was developed by psychologists Carlo DiClemente and James Proschaska in the early 1980s. They wanted to map out a framework for understanding change that could apply to many different types of behaviour and that could also be used by people from varying theoretical backgrounds. For this reason it became known as the 'transtheoretical approach'.

At the core of this model is a simple, and even obvious, idea: change doesn't happen all at once. Rather it occurs in increments or stages. You can apply this to pretty much any type of change. For example, if you're moving home, the actual moving is referred to as the Action Stage. But before you move, there's some planning that's needed – that's the Preparation Stage. And before you plan, you make a decision that comes after a period of thinking about it – that's the Contemplation Stage. There was also a time, further back, before you even started thinking about moving – that's the Pre-contemplation Stage. There are two other important stages too, but I'll come on to these later.

This model has been enthusiastically embraced by the addictions treatment field because it provides a useful map of where people can be in their journey of change. Some clients are in the action stage of taking steps and engaging in treatment. But many people in hospital for alcohol- or tobacco-related problems haven't reached the point of deciding to tackle their habits. Understanding about these different stages makes it easier to see what might be blocking change. Someone in the Preparation stage might want to change but not see how to. Someone in the Contemplation stage may be stuck in ambivalence, where part of them wants to change but another part isn't so sure.

This model can also be applied to the way we think and act in response to climate change and peak oil. Ten years ago, most people weren't even thinking about climate change. Now there's been a big shift; most people have moved at least into contemplation, and many into action. But people may be at different stages with different behaviours. They can be in the action stage of using low-energy light bulbs, but at the thinking stage when considering flying or car use. With peak oil, much of the public are still in the pre-contemplation stage of responding to this. There is a much lower level of public awareness about oil depletion. This is changing fast, though.

The other two stages are Relapse and Maintenance. With any change, movement can

The Stages of Change Model

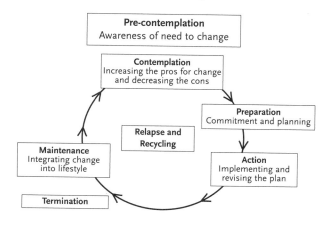

Figure 15. The Stages of Change Model. (After DiClemente, 2003)

be backward as well as forward. There may be initial good progress, but then people lose heart or become complacent, leading to a relapse back to a former stage. That's why the Maintenance stage of change is important – that we look at how to consolidate gains and keep going in the long term.

How do insights from the addictions field help the environmental movement understand how change happens?

A common idea in environmental campaigning is that if people only know how awful things are, then they will change. So the focus of many campaigns is on delivering information, often with disturbing graphic images and horror stories. Awareness-raising is of crucial importance – but you only have to look at a packet of cigarettes to see the limits of this approach. The information "Smoking Kills" in big letters isn't enough to discourage many smokers.

What the addictions field is good at is understanding and working with resistance to change. Approaches like Motivational Interviewing[4] have been developed as ways of working with people who have mixed feelings about change. There is massive resistance to tackling environmental issues, and we need to start being more creative in how we respond to this. There are important lessons here to learn from the addictions field.

To what extent is it possible to say that we are 'addicted to oil'?

As an addictions specialist, I'd say that industrialised societies are hooked on oil in a way that shows significant features of addiction. Many people accept this, including George W. Bush![5] But the term addiction can be a difficult one because it has no universally agreed definition. I still think it is a useful term though, as addictions refer to stuck patterns of behaviour that can be difficult to change even when we know they're causing harm. That's exactly what we've got with our current pattern of fossil-fuel use.

Many of my alcoholic clients find the term 'addiction' useful, because it helps explain why they find it so difficult to stop drinking. It needs more than just a conscious rational decision. Even when they do that, deeply ingrained habits can be hard to shift and temporary gains can be easily lost through relapse. But once they've accepted there's something called addiction in the way, they can give it their attention and learn strategies to tackle this.

Why might it be useful to say we are 'addicted to oil'?

In industrialised countries, a lifestyle that depends on very heavy oil use is seen as normal. The first stage in tackling a problem is to recognise it, and when we apply a term like 'addiction' to oil, it questions the way we use it. When looking at 'problematic substance use', it is useful to recognise three types of problem: hazardous use, harmful use and dependent use. Each of these can be applied to heavy oil consumption.

Hazardous use is when someone's consumption of a substance generates risks for the future. Many heavy-drinkers do not think of themselves as having a problem, but if they continue to drink at high levels, they increase the risk of medical complications. In a similar way, if we continue heavy use of fossil fuels like oil, we're likely to run into two main hazards – dangerous climate change and an energy famine when oil reserves run low. There's a saying in the addictions field: "If you carry on the way you're going, you'll end up where you're headed." Our current pattern of oil use is hazardous because of where it is taking us.

Harmful use is where consumption of a substance has already started causing problems. Climate change can be thought of as a toxic effect of heavy use of fossil fuels. In many parts of the world, people are already experiencing disturbed weather patterns. In Europe, heatwaves have killed thousands of people. In Africa, droughts have fuelled conflicts and famine. And in North America, an increase in the intensity of hurricanes has led to massive urban damage, most notably in New Orleans. While the future risks are much greater, the harm from climate change is already here.

If someone recognised their use of a substance was a threat to life, then in normal circumstances this would be enough to motivate change. But when someone is dependent on something, the idea of stopping use, or even reducing, is threatening. So in dependent use, someone may either block out information that suggests their favoured substance is harmful, or they may continue using it, even when they know it could be killing them. Dependent use is when someone is hooked liked this.

The value of recognising dependence is that it allows you to anticipate, and deal with, the additional obstacles to change this brings. Recognising oil dependence makes it easier to understand why it might be difficult to wean ourselves off our oil habit, while also pointing us towards proven strategies from the addictions field that might help us move forward.

So how can addictions treatment help?

Climate change tends to be thought of as an environmental issue, and peak oil as a resource issue: both might be seen as having distant causes that we can do little about. But oil dependence is to do with human behaviour; that's much closer to home and within our power to change. The Stages of Change model is useful

here because it maps out the journey needed for recovery.

The first stage of change is becoming aware of the problem. This starts us thinking about the issue, moving us into the Contemplation Stage. But it is easy to get stuck here if a conflict develops between the part of us that sees the need for change, and the part accustomed to using the substance and not wanting to go without it. Just think of all the things in your life that you appreciate that you wouldn't have if we didn't have oil. There are so many! And they become reasons for going slow in tackling the problem.

The approach of Motivational Interviewing developed as a way of dealing with such mixed feelings. By providing a listening space where someone can voice both their concerns and their resistances, ambivalence is brought into view where it can be dealt with. This helps people get clearer about what they really want, and so move into the next stages if they want to tackle the issue.

How can insights from addictions be utilised practically by Transition Initiatives?

I've boiled this down to three principles that are already being applied within the Transition movement:

a) Pay attention to the steps of change that happen inside people.

One of the lessons of addictions recovery is that information giving, by itself, isn't enough. In the Stages of Change model, becoming aware of an issue is only the first step; this moves someone from Pre-contemplation to Contemplation. It is easy to get stuck at this 'thinking about' stage, and this is where insights from the addictions field are helpful. By having 'Heart and Soul groups',

Transition Initiatives pay attention to the steps of change, and the blocks to change, that happen inside people. This allows them to address issues like motivation, resistance and ambivalence.

b) Create spaces for people to feel heard in making their own arguments for change.
A core insight of Motivational Interviewing (MI) is that when people make their own argument for change, they talk themselves into tackling an issue. Rather than trying to persuade people, the focus of MI is on creating a listening space that supports people to express their hopes and concerns. This is a way of cultivating the motivation needed to work through ambivalence and move through resistance.

Most political meetings have an active speaker talking to a relatively passive audience. A motivational interviewing approach might also add an opportunity for the audience to feel heard in making their own arguments for change. Several Transition events have done this through the use of paired listening exercises. At the launch of Transition Initiatives in Totnes, Lewes and Bristol, hundreds of people divided into pairs, with one as listener, the other as speaker. The speaker had two minutes of listening time for each of the following open sentences.

> "When I think about Peak Oil and Climate Change, concerns I have include . . ."

> "My positive vision for what I'd like to see happen in this town/city is . . ."

> "Steps I can take to help make this happen include . . ."

The listener's role was just to give full attention to what was being said. The roles swapped after the three sentences, so that everyone got a chance to speak. This process took about twenty minutes, and visibly raised the level of energy and enthusiasm amongst those present. After the Bristol meeting, one participant said: "When we spoke in pairs, something happened in the room. That was when we became a community."

When we express our concerns, we talk ourselves into addressing them. When we give voice to our visions, we identify the destinations we want to move towards. And by describing the steps we can take, we prepare ourselves for action. This simple tool is an example of a 'motivational nudge'; it can help provoke the inner steps of change.

c) If a change seems too difficult, have a preparation stage for training ourselves.
Changing an addictive behaviour can be so difficult that people sometimes give up, believing it to be impossible. In my clinical work I've found it helpful to think of recovery as a journey that may move through a 'phase of disbelief'. I draw inspiration from adventure stories that often begin from a similar place of gloom. When the main characters rise to the challenge and begin looking for a way, they are more likely to find one. The quest for a way forward usually involves seeking out mentors and guides, who pass on the skills and insights needed to turn the impossibility around. By including a Preparation stage, the Stages of Change model offers an alternative to giving up when challenges like peak oil or climate change seem too difficult to address. The Preparation phase is where we train ourselves to strengthen our capacity to respond.

Transition Initiatives don't just involve telling people about the problem and campaigning. They also involve practical training in the skills needed for a post-oil society. But as well as practical training, psychological training is also needed in how to cultivate positive visions and find ways of dealing with inner 'dreamblockers' like fear, cynicism and disbelief.

Disbelief can be challenged by seeking out inspiring examples: addictions recovery, adventure stories and Cuba's bounce-back from energy famine are reference points that support the possibility of turnarounds from oil-dependence. In my book *Find Your Power* I offer a toolkit of strategies for turning around feelings of impossibility and finding ways through inner blocks to change. Such inner training is part of the preparation needed for creative transition out of the oil age.

What strengths might the integration of these tools and insights add to Transition Initiatives?

Environmental campaigns tend to focus on awareness and action. But between these there's a series of internal steps, and change can become blocked at any one of these. Transition Initiatives are strengthened when they take account of both the inner and outer dimensions of change. Without this, when we encounter resistance to change we're in danger of falling into complaint and blame, rather than developing understanding and insightful responses.

There is a close parallel to what happens when someone has a drinking problem. Nagging responses in relatives are understandable, but they can deepen the resistance to change. We need to accept that when people are dependent on substances, as we are with oil, there are resistances to change that we need to take into account. The addictions field has been working with such resistance for decades. Models have evolved for understanding and working with blocks to change. Effective tools have been developed. The challenge we face is about transition and using the tools and insights from one field in another.

The FRAMES model

One model from the addictions field that I have found to be particularly useful, and which offers a way of pulling together the threads of this chapter – indeed of the book thus far – is something known as the FRAMES model, devised by Miller and Sanchez.[6] In the context of this book, the FRAMES model offers a template for how we can apply insights from addiction to practical responses to energy descent. The overlapping of these two fields is very exciting. In essence, the FRAMES model comprises six elements commonly included in brief interventions to addiction that have shown to be particularly effective. We could think of them as being best practice for responding to addiction.

The acronym FRAMES stands for:

- **Feedback**
- **Responsibility**
- **Advice**
- **Menu of options**
- **Empathy**
- **Self-efficacy**

(Not in any kind of chronological order).

Feedback of personal risk or impairment

In the drug and alcohol field, this relates to offering the client an honest assessment of their addiction problem and its potential consequences in order to raise awareness of the problem. In relation to peak oil, many groups begin by showing the film *The End of Suburbia*, a frank assessment of the peak oil challenge.[7] An essential element of initiating successful responses is making the level of the problem clear in stark terms. There is clearly a balance to be struck between the potential sense of disempowerment and trauma that may be generated and a positive solutions-focused programme.

Tools for Transition No. 3: The humble potato becomes a tool for breaking our oil addiction: an exercise

This activity evolved from an exercise called 'bunyips' in Skye and Robin Clayfield's *Manual of Teaching Permaculture Creatively*, but is an evolution of my own devising. It is a powerful tool for giving people permission to explore how they might set about lessening their oil dependency. I have found it to be very effective with many different groups.

First Step

I use this exercise in weeks 9-10 of the ten-week Skilling Up for Powerdown course. It is a very useful tool for pulling all of these threads together, but it does require a degree of underpinning knowledge and enthusiasm, so is best not done too soon.

Second Step

Link our relationship to energy use and consumerism in general as being like an addiction, and then give the students a sheet which reads as follows:

> "My Twelve Step programme for reducing my oil dependency. In order to make my life less reliant on the unreliable, I pledge to myself to strive towards the following 12 goals over the next 6 months."

Tell them that by next week's class they are to have filled out the sheet, with twelve actions which are achievable, practical but ambitious.

Third Step

Get people into pairs to do a five-minute each-way 'Think and Listen' (see page 165) on the

question "when you think about making practical steps to make your life less oil dependent, what are the obstacles you put in your way of doing that? What are the voices within you that block you taking firm steps?" Once they have discussed this, ask them to reflect in silence on their own for a minute or so on the question "what might the antidotes to this be? If you could name the qualities you would need to overcome these obstacles, what would they be?"

Fourth Step

Ask them if they were able to personify these qualities and strengths in a Superhero, what would his/her name be, and what would his/her powers be? Ask them to think about that for a while and then hand out to each person four potatoes and a few cocktail sticks.

Give them twenty minutes to make their Superhero, and tell them that they will be expected to introduce their hero to the rest of the group, its name and its powers. They can also use anything else in their model that they find lying around.

Once they have finished their models, break them into groups of five or six, and let each person introduce their Hero. Tell them that they are to take their Hero home, put him somewhere where they will need his/her powers while they are writing their Twelve Step Programme, but they are NOT to tell anyone in their house what it is, or else it loses its powers.

Fifth Step

The following week, when everyone has brought their completed sheets along, I start by giving a talk about the Transition concept and energy descent (this would be specific to what you are doing in your Transition initiative).

Then arrange them into two groups of an equal amount of people. One group sit in a circle in the middle facing outwards, the other half sit outside them and face in. Then they do a process based on Speed Dating. They have four minutes in which to tell the person facing them about their Twelve Step Programme. Every four minutes you tell the outer circle to move one seat to the right. In this way everyone gets to hear half of the group's plans.

Sixth Step

Divide them into groups of five or six and get them to sit around tables. This final session is based on the World Café approach (see pp.184-6).

One person is the Table Host who takes notes, and everyone else discusses the issue. After 10 minutes everyone gets up and goes off to find another table to continue the conversation. The three questions asked are:

"How can Transition Town Totnes (replace 'Totnes' with the name of your town or area) help you to achieve your Twelve Step Programme?"

"How can you help Transition Town Totnes/your town?" and

"How can you help each other achieve your Twelve Step Programmes?"

Seventh Step

Once these three conversations have taken place and the notes taken, tell them that when the course runs again (you'll have to adapt this bit to suit your situation) they will be asked to come in and mentor the people on the course, to tell them about their Twelve Steps and what they have done with them since they did the course.

Reflections

This exercise really gets people engaged with the idea of grounding these things in their lives. Steps they choose to make range from changing their light bulbs to doing a course, or from getting involved in their local Transition process to selling their cars. The exercise with the potatoes is very powerful. If you asked them to model their Superheroes in clay, the models would become very precious and overworked. If you ask them to make them in knobbly potatoes and cocktail sticks, it is impossible to make them in any way that they do not look ridiculous and daft, yet at the same time people imbue their characters with all the powers of their Heroes.

Emphasis on personal responsibility for change

For Miller and Sanchez, this relates to making an alcoholic/addict aware of the degree of personal responsibility that breaking the addiction will require. In the energy descent field, this relates to emphasising that the creation of the problems of peak oil and climate change is the result of many individual actions, and that the solution requires taking responsibility for these actions. Clearly, a response akin to a 'wartime mobilisation' will require the large majority of people taking on some of this responsibility. The emphasis here is on individual responsibility and choice, rather than merely telling people what they should do.

Clear advice to change

Clear advice needs to be offered to break an addictive pattern. Advice has been shown to be effective, but it needs to be given as a recommendation not as a prescription. It can come in two forms: firstly advice to individuals for modifying their own lifestyles, such as energy-efficiency advice, and secondly, as community-scale strategies for energy descent. Indeed, one could see an Energy Descent Plan as being clear advice to change on a community scale, setting out a plan for responding to what is rapidly becoming seen as a disastrous addiction with potentially catastrophic results.

A menu of alternative change options

In order to feel ownership and a sense of responsibility for an Energy Descent Action Plan, people need to feel that they have explored the alternatives. To arrive at the recommendations this plan would contain requires a process of exploring what the different options might be. Here the use of scenario planning is very useful, as it enables people to project forward and explore different possible outcomes (some of these scenarios were explored in Part One). Other tools that are useful here are visioning and backcasting,[8] and one of the forms of this being explored by Transition Initiatives is the Transition Tales project (see pp.118 and 200) being developed in Totnes, which invites people to tell stories through a variety of media, making a power-down future feel like a tangible reality.

Therapeutic empathy as a counselling style

In the field of addictions, the idea that aggressive, authoritarian or coercive approaches are effective tools is increasingly being discredited. What is now accepted as better practice is for the role of the counsellor to be supportive, friendly, encouraging and empathetic. Similarly, any approach that seeks to engage a significant proportion of the population in responses to energy descent has to skilfully engage with people and instill a sense of optimism regarding the possibility of change, rather than berating them for their planet-wrecking ways. This creation of a sense of embarking on a collective journey which Chris Johnstone refers to may well be key to this.

This principle also implies that the dialogue is a two-way process, that the person imparting the information is open to receiving information as well as giving it. Rather than telling people what they should be thinking and/or doing, an empathic approach seeks to engage as well as educate.

Enhancement of client self-efficacy or optimism

This is key to the success of this process. The term self-efficacy refers to an individual's estimate or personal judgement of his or her own ability to succeed in reaching a specific goal, such as giving up alcohol or reducing their degree of oil dependency. Building this sense of 'can do' is essential in catalyzing change on the scale we are talking about. You will see in Part Three some of the ways in which Transition Initiatives are building this sense of optimism and working, through various approaches, to build self-efficacy – a community-wide belief that we can actually do this.

This is one of the key areas that goes beyond the familiar approach to environmental campaigning with which we have become most familiar, that is about disseminating information. There is a real challenge too, in terms of how to create that sense of self-efficacy in diverse populations and how to design an approach that engages the diverse range of people that make up our communities.

IS 'PEAK' OIL THE MOST USEFUL WAY OF LOOKING AT THIS?

When we look at the standard Hubbert curve, we see a mountain: a rise followed by a fall, an ascent followed by a descent. There is a sense that we have reached the peak and that now we have to grit our teeth for the long journey home, akin to an overexcited child at a birthday party being told it is time to go home. Perhaps the sense that we need to instil could come from turning this much-viewed graphic upside down. We might more usefully use the term 'trough oil'.

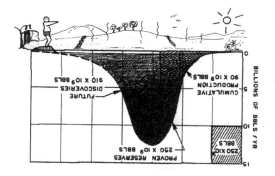

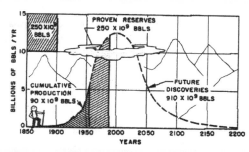

Rather than a mountain, we could view the fossil fuel age as a fetid lagoon into which we have dived. We had been told that great fortunes lay buried at the bottom of the lagoon if only we were able to dive deeply enough to find them. As time has passed we have dived deeper and deeper, into thicker, blacker, stickier liquid, and now we find ourselves hitting against the bottom, pushing our endurance to the extreme, surrounded by revolting sticky tar sands – the scrapings of the fossil fuel barrel. We can just about see distant

sunlight still glinting through the liquid above us, and our desperate urge to fill our lungs begins to propel us back upwards, striving for oxygen.

Rather than being dragged every step of the way, we propel ourselves with focused urgency towards sunlight and fresh air. Viewed like this, the race for a decarbonised, fossil fuel-free world becomes an instinctive rush to mass self-preservation, a collective abandonment of a way of living that no longer makes us happy, driven by the urge to fill our lungs with something as yet not completely defined, yet which we instinctively know will make us happier than what we have now. Perhaps arriving in a powered-down world will have the same sense of nourishment and elation as finally breaking through the surface and filling our lungs with fresh morning air, marvelling once again at the beauty around us and the joy of being alive.

Chapter 7

Harnessing the power of a positive vision

THE OPTIMISM/PESSIMISM TRAP

"I've found myself bouncing back and forth between optimism and pessimism. "Things are going to work out well." Or: "There's going to be real disaster!" It's been really exhausting.

But lately something's changing about all this.

I've begun to notice how the whole optimism/ pessimism dichotomy is a death trap for my aliveness and attention. I watch myself acting as if my sense of what might happen is a description of reality. And what I notice is this: whether I expect the best or the worst, my expectations interfere with my will to act.

That's so important I'm going to repeat it. Whether I expect the best or the worst, my expectations interfere with my will to act.

I've started viewing both optimism and pessimism as spectator sports, as forms of disengagement masquerading as involvement. Both optimism and pessimism trick me into judging life and betting on the odds, rather than diving into life with my whole self, with my full co-creative energy. I think the emerging crises call us to

(continued on facing page)

It is one thing to campaign against climate change and quite another to paint a compelling and engaging vision of a post-carbon world in such a way as to enthuse others to embark on a journey towards it. We are only just beginning to scratch the surface of the power of a positive vision of an abundant future: one which is energy-lean, time-rich, less stressful, healthier and happier. Being able to associate images and a clear vision with how a powered-down future might be is essential.

I like to use the analogy of inviting a reluctant friend to join you on holiday. If you can passionately and poetically paint a mental picture of the beach, the sunset and the candle-lit taverna by the sea, they will be more likely to come. Environmentalists have often been guilty of presenting people with a mental image of the world's least desirable holiday destination – some seedy bed and breakfast near Torquay, with nylon sheets, cold tea and soggy toast – and expecting them to get excited about the prospect of NOT going there. The logic and the psychology are all wrong.

Why visions work

My sense is that creating a vision works in many interrelated ways. Tom Atlee writes of creating what he calls an "alternative story field".[1] This in essence is creating new myths and stories that begin to formulate what a desir-able sustainable world might look like. He talks of the potential power of bringing together activists, creative writers and journalists to form 'think-tanks' that create new stories for our times. When we start doing Energy Descent work, we should be looking to draw in the novelists, poets, artists, and storytellers. The telling of new stories is central. In Totnes we have started to do this with our Transition Tales initiative, which aims to get people writing stories from different points during Totnes's transition, as newspaper articles, stories, or agony aunt columns. Some of these appear later in this chapter. Such stories can come in all sorts of forms.

The pilot Totnes Pound that Transition Town Totnes (TTT) ran until June 2007 was also an example of this. People were able to hold in their hands a tangible, beautiful and spendable banknote. It told a new story about money, about its possibilities and about their community. The concept of telling new stories was also raised at the Official Unleashing of TTT in September 2006, when Chris Johnstone said:

"Totnes has an opportunity here to be ground-breaking internationally. Maybe in 400 years time, if humanity finds a way through, they will look back at this time at the beginning of the 21st century as a crucial time, as the last decade of the Oil Age. Maybe they will tell stories about

SUSSEX EXPRESS

1 February 2017

TRANSITION TOWN LEWES: A VISION OF THE FUTURE

By Mavis Happen

Last Saturday Lewes celebrated winning the 2017 Synergen Town of the Year award with a festival at the North Street Centre that included a Southern Solar[2] disco and a local produce feast. Lewes was among 965 entrants for the award, for the 'most creative energy descent programme[3] and improvement of living standards'. The judges were particularly impressed with the depth and effectiveness of the town's '2020 Vision', which was adopted unanimously by the Council in 2008,[4] along with the Climate Change Strategy, written by the Sustainability Team in 2006 and Zero Waste Target[5] reached in 2012.

'We're really proud of our achievements,' said Cllr Billie Turner of the transition committee. 'It's been a lot of work, but exhilarating. Of course, the floods of 2000 and 2008, and the hurricane of 2010 really helped focus our minds. That and oil prices hitting $350 a barrel a while back.'

Jewel in the crown of the town is the North Street Centre, five hectares of land at the heart of Lewes's 2020 Vision. Youth worker Toma Stevenson pointed out the riverside ecovillage. 'They're on stilts to be flood-flexible. All 200 houses sold really fast. It was the first development in the UK to be a fossil-fuel neutral build,[6] and because half were affordable housing, many of the people living there work in the North Street Centre too.' There's now a ten-year waiting list for the car-free ecovillage:[7] a recent survey put residents at the top of the UN Happiness Index.[8]

'The biomass plant over there was put in around 2015 to turn fast-growing willow from the floodplains into electricity,' Toma continued. 'That willow soaks up the flash floods we were starting to get a few years ago. The launch of the community wind farm (Lesco – Lewes Energy Supply Company[9]) made us one of the first towns to export electricity to the national grid.'

'Over there by Furniture Now,[10] Plumpton has an urban agriculture training centre. Those raised beds have been producing 40 tons of food per hectare for 8 years now. Most of it goes to the big weekly riverside market and the dozens of local produce shops around town, and any surplus goes to Lewes Preserves. This, along with the twenty-odd CSA (Community Supported Agriculture[11]) farms in the area, Lewes Allotments 2020, and the council's Home Grown initiative, means that 75% of Lewes's fruit and vegetables are grown[12] within 5 miles of the town, just as it was this time last century. The jury is still out as to whether the supermarkets in the town will win their legal action against smallholders preferentially supplying the local markets.

The North Street Centre has become the transport hub for Lewes, linking the biodiesel bus station, cycle station,[13] working horse stables and car club depot. Car ownership[14] in Lewes is now well below the national average ratio of 1:4.

'Ten years ago, we had had a developer pitching to build intensive high-rise housing, car parks and a chain store centre in North Street. Our progressive council ran a weekend Appreciative Enquiry[15] summit for that site, and found that people were very concerned about the environment. That form of effective consultation resulted in our 2020 North Street vision for a transition future with low-impact development that allowed the possibility of flooding.

'Fortunately, our District Council unanimously adopted that 2020 Transition Strategy for North Street at the end of 2007. Things could have looked very different!'

transcend such false end-games like optimism and pessimism. I think they call us to act like a spiritually healthy person who has just learned they have heart disease: We can use each dire prognosis as a stimulant for reaching more deeply into life and co-creating positive change.

And so I've come to conclude that all the predictions -- both good and bad -- tell us absolutely nothing about what is possible. Trends and events only relate to what is probable. Probabilities are abstractions. Possibilities are the stuff of life, visions to act upon, doors to walk through. Pessimism and optimism are both distractions from living life fully."

– Tom Atlee,
'Crisis Fatigue and the Co-creation of Positive Possibilities',
Co-Intelligence Institute,
www.co-intelligence.org

Tools for Transition No. 4: The Board Game – a tool for identifying your vision and making it real *by Naresh Giangrande*

The Board Game is a powerful tool for the identification and design of a project's many stages, developed by John Croft of the Gaia Foundation of Western Australia (see Appendix 2). It also creates a checklist to which one can return as the project progresses, to get a sense of how it is progressing. At an early stage in the Transition Initiative, the initial group of people who are passionate about the process comes together. The ideal suggested circle size is 6-8 people – with more than that, break into groups and put the objectives together at the end.

Step One: What Do We Want To Achieve?

Each person is asked the following question in turn: "What would this project need to achieve to make your participation 100% worthwhile?"

Croft suggests using a talking-stick or similar, so that people are listened to, and then to keep going round the circle until everyone has their full set of objectives listed. When you are satisfied you pass – and it may take three or four rounds until everyone passes. There is no evaluation, competition or criticism at this stage – just go on including all the objectives. When you have finished you have a list of objectives for the project, and the aim of the Board Game is to design a way to complete 100% of them.

Step Two: Creating the Board Game No. 1

On a flip-chart-sized piece of paper draw a circle at the top labelled "Start" (where you are now and what the project is) and one at the bottom labelled "Finish" (where you want to get to and what, for the group, will constitute success). A project can be planned and created up to the point of its official launch, or to the point when the last champagne cork has been popped and the last penny spent and there is nothing more to do. It depends on the project and on your aims (each phase can also have its own board game).

Now invite the group to brainstorm the steps and tasks that will be needed to complete the project's objectives. Central to Croft's model is that any project has four key stages, dreaming, planning, doing and celebrating. Divide your sheet into these four sections and as the ideas start to flow from your brainstorming, place them in whichever of the four sections is

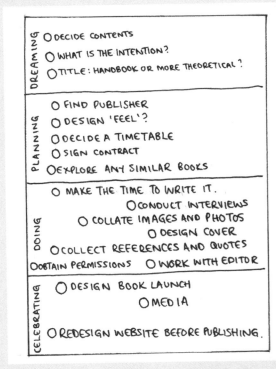

Figure 16. A Board Game identifying the process of producing this book, showing various stages of the process.

most appropriate. Croft suggests that for those less experienced with this process you start by making a list of all the tasks and then labelling them 1 / 2 / 3 / 4 (or more than that if people see more than one) for each of the four stages.

When you have your list of tasks you can get a sense of the strengths and weaknesses of the group and of the project, whether you have mainly planners, dreamers, doers, or celebrators. Then give each element a circle.

Step Three: Creating the Board Game No. 2

Now add lines to start joining up your tasks. Where do you start, and what logically follows on from that? Every task should have at least one input and output – if you can't find a way to con-

nect tasks, then there is a step in the process that is missing which you have forgotten about or not thought of, so where appropriate add them in.

Step Four: Using Your Board Game

When everything is connected up you have the plan for your project. Then ask each person in the group to identify:

- which task/s they feel passionately about – add their initials in one colour

- which tasks make them feel scared witless, or really afraid of taking on – add initials in 2nd colour

- which tasks they feel competent to do – add initials in 3rd colour.

These are the people who should do the tasks. People taking on fearful tasks will need support – someone who knows what they're doing. Tasks with lots of inputs and outputs are key, and you need to make sure they are resourced properly.

Step Five: Your Board Game as an Ongoing Tool

As soon as you have completed your Board Game, go through the Board to see where your project is at the moment (i.e. some of it will have already started). Then:

- cross-hatch the circle of any task you have started

- fully colour in the circle of any task you have completed.

This Board Game is the agenda for every ensuing meeting, and an ongoing record of what's happening. Keep hatching and colouring in as your project progresses, and make sure to celebrate each time you do that. If an individual task is complex it may require its own Board Game – with the understanding again that within each task are the four steps of the process.

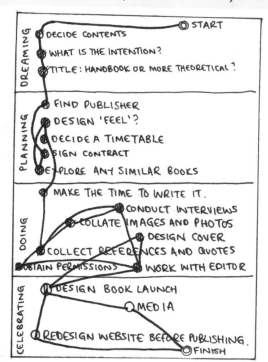

Figure 17. The same Board Game with the elements joined up, identifying the order in which the project should proceed.

what happened in Totnes. Maybe this evening will be something that is the beginning of one of those stories. If you look ahead at the future, there are gloomy possibilities, but there are also inspiring possibilities, and you are part of an inspiring possibility by being here tonight." [16]

The tool of visioning offers a powerful new approach for environmental campaigners. We have become so accustomed to campaigning against things that we have lost sight of where it is we want to go. One of the best examples of this recently was provided by Transition Town Lewes, which when confronted by a local developer who wanted to develop a key part of the town, responded not with protests and petitions, but with a vision – the fictitious newspaper article on page 95.

This is a great example of what Atlee calls 'imagineering', and the creation of what he terms *The Ecotopian Grapevine Gazette* which he describes as "contain(ing) imaginary news stories about events or innovators that had not happened yet, but which I and others wanted to have happen, written as if they had happened. At the end of each article, I put the contact name of someone readers could call and participate in making that story a reality." [17]

Peter Russell, the physicist and writer, describes a collective vision in terms of a Strange Attractor, as described in chaos theory. In effect it is like throwing a whirlpool in front of you which then draws you towards it. It has an energy, it is dynamic. He adds:

"There's something deeper which I can't really explain, but when there is a vision, it's somehow not just a motivation, but somehow the psyche gets involved in a way that seems to interact with the world in a way that makes it easier for things to actually happen, things seem to fall in place. I can't explain that rationally, but it's something

that people notice time and time again. If you've got a strong vision of where you're going – it's as if the world seems to want to support that vision. It just seems to do it." [18]

Visioning in this way has the added benefit of counteracting despondency. Climate change and peak oil can be terrifying, bewildering or seen as inevitably catastrophic. James Lovelock's recent book *The Revenge of Gaia*,[19] with its paperback edition featuring a cover like a 1950s B-Movie horror film, and websites such as www.dieoff.org set out scenarios so grim that most people simply switch off; they don't want to engage with them. I am aware that being one of those people who can read a desperately depressing book about peak oil and societal collapse and draw from it the inspiration and motivation to do something practical puts me in an extremely small minority.

As a species with the creativity, adaptability and opposable thumbs that enabled us to create an Oil Age in the first place, we can be pretty certain that there will be life beyond it. Similarly, we may be able to prevent the worst excesses of climate change, and indeed the measures needed would almost certainly make the world a far better place. However, the point is that the world and our lifestyles will look very different from the present. It is worth remembering that it takes a lot of cheap energy to maintain the levels of social inequality we see today, the levels of obesity, the record levels of indebtedness, the high levels of car use and alienating urban landscapes. Only a culture awash with cheap oil could become de-skilled on the monumental scale that we have, to the extent that some young people I have met are lucky to emerge from cutting a slice of bread with all their fingers intact. It is no exaggeration to say that we in the West are the single most

magazine, and thinking that holiday trips to the moon would be on the cards for my adulthood. I'm still waiting, and, however much Richard Branson might think holidays in space might be possible, I'm not holding my breath.

Captain Future – the wizard of science

On the left is a picture of Captain Future, the 'man of tomorrow', published in the US between 1940 and 1951, as a kind of long-running 'space opera'. Captain Future is the *alter ego* of Curtis Newton, a space-traveller and scientist for whom life is a non-stop whirl of vanquishing supervillains, tackling alien menaces and generally being a good guy in a universe of oddly

useless generation (in term of practical skills) to which this planet has ever played host. However, the first step to the creation of a localised, low-energy-abundant future is actually visioning its possibility.

I can, of course, lay no claim to the idea of visioning the future as being unique to this book. Throughout history, humankind has always created visions of how the future might be – living in space stations, flying to work in our own flying saucer, going on holiday to the moon, for example. They rarely ever come to pass, though, usually owing to our not taking into account, among other things, the amount of energy that sustaining such visions would require. I remember as a child lying on the floor of my bedroom reading *Look and Learn*

shaped baddies. The stories begin in 1990(!) when scientists Roger and Elaine Newton and their brilliant scientific colleague Simon Wright move their laboratory to the Moon to enable them to continue their research in peace and quiet. Wright is very old and unwell, so in order to be able to keep working he transplants his brain into a clear perspex box, with wires on it which, when attached to someone's head (see previous page), enables them to have Wright's thoughts (a very early but somewhat more revolting version of the internet).

Newton and Wright create various things, including a robot called Grag and a shape-shifting android called Otho. Their lunar scientific idyll is cruelly spoilt when the baddie, the evil and utterly dastardly Victor Kaslan, arrives and

kills them all, apart from their son, who is then brought up (somewhat unconventionally) on the Moon, by a robot, an android and a brain in a box. It is a childhood, however, on which he thrives, growing up to be an incredible scientist and athlete for whom no challenge is too much.

Captain Future had a penchant for hovering space boots (which disappointingly never came to pass) and helmets that miraculously never steamed up. He also appears to have relished squeezing women into absurdly small rockets that would panic even the most agoraphobic flyer (see left).

Of course, much of Captain Future's vision of the future didn't come to pass. We don't have household robots, chasing villains through space is something that thankfully we will

Tools for Transition No. 5: Post-peak tour guides

This is a visioning exercise, one I use most often on permaculture courses when teaching about urban permaculture, usually about two-thirds of the way through the course. The scenario is that it is 2030, and that the town/city/village you are in successfully made the Transition to a lower energy, more localised model. As such, it is now an exemplar for the rest of the world similarly engaged in this process. People come from far and wide to draw inspiration from what has been achieved. Their job is to act as tour guides.

Take the group to a central point in a housing estate. Divide them into groups of four or five, and give each a subject area (housing, food, energy, waste etc.). Tell them they have twenty minutes to design a ten-minute tour for the rest of the group which departs from and returns to this central point. They are to introduce us to all the fantastic developments that have made this place so famous.

When they lead their tour, encourage them to really help everyone else to see and feel what we are 'looking' at, to tell stories and to really bring it to life. Once all the tours are complete, it is very useful to then reflect on whether the current design facilitates or inhibits the place's ability to adopt these ideas. For example, if the houses had all been designed to face south to start with, how much difference would that have made?

One small caveat: do be mindful of the fact that you are walking around in other people's space, and make sure that the students are respectful of that. It is not about criticising how things are done. I once did this exercise in a town north of Cork in Ireland on a Saturday afternoon.

One of the participants was a very loud and outspoken Australian lady, who walked around, in earshot of the residents who were washing their cars and clipping their hedges, proclaiming "I mean, how do people live like this? Really, it's all about education", and so on. I had to take her to one side and have a quiet word. Also, groups of people wandering around with clipboards can make people nervous, thinking you are developers planning a motorway through their gardens, so be prepared to explain to people what you are doing. This is a very popular exercise, one that really grounds the concept of visioning in an everyday setting.

never get to do, and our future is likely to be far more terrestrial, far less well-travelled, and with the robots limited to the inside of car factories. It will be, however, ultimately preferable. Nonchalantly zapping enormous marshmallow monsters on the Moon (see page 100) is not something we will be called upon to do. Captain Future's scientific brilliance and his powerful physique are something we might find useful, however.

Visions of abundance

(see page 100)

Not all visions of the future have to be quite as absurd as that of Captain Future. I have, over the last year or so, become what one might call a voracious 'vision-harvester'. I am liable to ask the various ecological thinkers and practitioners I bump into what their visions are for a powered-down future. Their answers are intriguing. For some, such as Stephan Harding, author of *Animate Earth*,[20] a central part of his vision is the return of the role of wilderness in our lives, our decreased ecological footprint having allowed nature to reclaim some of the space around us.

"I would know," he told me, "that I could put on my rucksack and walk out of the village (where I live) and into the forest and be really in the wild country for days and days if I wanted to. . . . My vision would be for an interconnected network of ecovillages, with lots of wild countryside in between, but also some lovely small cities where there would be theatre, culture, museums and good libraries, and good coffee shops, gorgeous organic architecture."[21] The psychological and cultural effects of this reconnection with Gaia would be, he argued, hugely beneficial.

For Brian Goodwin, author of *Nature's Due*,[22] a powered-down future is one where humanity becomes, as he puts it, "largely invisible"; that is, more blended into and in tune with our natural surroundings. He told me "I'm not talking about a Rousseau 'back to nature'; I'm talking about using appropriate technology, natural materials and energy to achieve a lifestyle in which we blend with the natural world. We will have learned how to live in a way that other species have, and therefore have reduced our footprint, decreasing it dramatically to the point where we are one among many instead of an absolutely dominant species."[23]

Systems thinker Fritjof Capra's vision of 2030 is one where the ecological principle of community has become the central organising factor for society. Taking nature as the model, he told me, would mean that "we would have patterned our communities after . . . natural communities, which means that we would use solar energy as our main energy source, augmenting it with wind, biomass, and so on. We would have arranged our industries and our systems of production in such a way that matter cycles continuously, that all materials cycle between producers and consumers. We would grow our food organically, and we would shorten the distance between the farm and the table, producing food mainly locally.

"All of this would combine to create a world that has dramatically reduced pollution, where climate change has been brought under control, where there would be plenty of jobs, because these various designs are labour-intensive, and as an overall effect there will be no waste, and the quality of life would increase dramatically."[24]

For Meg Wheatley, author of *Leadership and the New Science*, this kind of visioning is not hard, as she recognises the qualities of this future and the relationships we would have with those around us, in existing communities she has already spent time in. She identifies

"If you have built castles in the air, your work need not be lost, that is where they should be. Now put foundations under them."

– Henry David Thoreau

these as being communities "where you recognise that you're all working for the same values, for a shared vision, for similar goals, and you're not working at odds. You don't feel polarised, you don't feel afraid of truthful conversations, and you don't retreat from each other, whether it's because of conflict, or just because I have no patience for what you think and contrasting it with what's so prevalent right now." [25]

For Tony Juniper, former executive director of Friends of the Earth, the principal noticeable difference would be that it would be quieter, and people would be in less of a rush. "There would be more sounds of people and less sounds of machines," he told me, "because communities would have been rebuilt and there would be people back in the streets once more, meeting each other rather than exchanging abuse through their car windscreens!"

According to Juniper's vision, the improved quality of life would be tangible. "It would smell fresher, there would be less pollution, less noise as well. . . . There would be more bicycles, more birdsong because the pollution that has been associated with industrial agriculture would have declined, there would be more organic methods, so there would be more wildlife back in the countryside and the cities." [26]

Whatever happens, it is clear that what will happen over the next twenty years is almost unimaginable. When I asked Dennis Meadows, one of the co-authors of the *Limits to Growth* series of books, he said: "If you think about the degree of change you saw in the last 100 years – social, technical, cultural, political, environmental, all those changes – it's less than what you'll see in the next twenty years." [27] These are extraordinary times.

Chapter 8

A vision for 2030:
looking back over the transition

If I were to wake up in the UK of twenty years hence, which had successfully undergone a transition of the kind we are talking about, life would be very different from the present. It would have emerged from some very difficult times into a more settled place. It would be far more locally oriented than the present, and we would have much less reason to travel. Let's take a look at how 2030 might look from the perspective of someone in 2030 looking back. I have illustrated this below with some newspaper articles from various points in the future.

Food and farming

Farming has experienced a remarkable transformation, undergoing a renaissance that few in 2008 could have thought possible. About twenty years ago, rising oil prices, international climate change agreements, and the findings of the Royal Commission on Food Security,[1] made the UK Government reconsider its commitment to the World Trade Organisation's pro-globalisation, liberalised, unrestricted free-trade approach, leading to its re-prioritising national food security above international trade. Also, local authorities across the country made local food procurement a policy priority, thereby kick-starting a rapid expansion in the market for local food. Farms are now highly diversified, producing more than just food, and are also providers of local-scale renewable energy, building materials, and organically grown medicinal plants, among other things.

Rising natural gas prices and disruptions to supply exposed the vulnerability of UK farming's dependence on nitrogen fertiliser (which was made from natural gas). The building of organic matter in soils due to their increased ability to lock up carbon is now a priority, a key aspect of the Government's carbon reduction strategy. The integration of perennial tree crops is a central feature of agriculture, both for their crop yields and for their carbon-sequestering abilities. Stands of specially bred varieties of walnut, sweet chestnut and hazel have been integrated into most farms, offering protein-rich annual crops for a variety of food uses as well as for producing oils for biodiesel for local consumption.[2] With the changes in climate, a wider range of tree crops is now grown, as well as vines and other perennials.

Farming has learned to compensate for its reduced oil consumption through the partial reintegration of working horses,[3] alongside locally produced biofuel-powered machinery, and by employing more people. The average farm size is now much smaller than in 2008 and the countryside is substantially more populated. Farms are now host to a diversity of enterprises, not just food production: some now produce materials needed by a building industry now using more local building materials, such as clay plasters,[4] cob[5] and hemp/lime

TOTNES TIMES

Incorporating the — Totnes Gazette

Wednesday, April 18, 2015   Established 1860   1 Totnes Pound

INSIDE THIS WEEK

Natural Building A level Success!
page 3

Local Biogas Co-op Opens
pages 13-15

John Davies: the reed bed king!
page 6

Chickens and Vegetables Bring New Life to Totnes!

"I Come Here Every Day – It's so Relaxing"

HUNDREDS TURNED OUT LAST WEDNESDAY for the official opening of the Transition Land Trust's market garden, known as New Gill's Nursery, on the site of a former car park in the middle of Totnes. Two years ago, the Town Council voted by a narrow margin to allow a change of use for the car park. Car ownership had fallen significantly since 2012 to the point where it was felt that for the town to lose one of its four car parks would have little impact on the viability of local businesses and indeed could become an attraction for those travelling to the town by boat and via the recently improved railway service. At the same time, rising fuel prices have made local food production increasingly economically viable.

by Nora Livingston

John Wheatcroft, manager of the garden, explains what has been created. "We have half an acre of intensive vegetable production", he explains. "They are all grown in 3-feet-high raised beds, allowing our disabled gardeners to access the beds. Through the use of our two glasshouses, we grow salads and greens all year round, and the design of the garden to cleverly integrate our flock of hens means that we have no slug problem."

Sarah Bishop, 8, was enjoying the gardens. "I like the chickens best", she said. "They're so funny. I bring the crusts for my lunchbox here after school, and I've learned

Sarah Bishop enthusiastically attending to her tomato plants at New Gill's Nursery in Totnes.

to like loads of interesting kinds of salad. Last week they showed me some flowers you can actually eat!"

The garden was started with money raised through a community Land Trust as well as a grant from the National Lottery. The ribbon was cut by winner of last year's Celebrity Love Allotment, Letitia Lloyd, who praised the garden as a model for towns and cities across the UK. "Beats a car park any day!", she told the assembled crowd.

"Can gardening seriously hope to replace farming as a major food source for urban people? An answer to this is suggested in an interesting report called 'The Garden Controversy', published in 1956. The authors found that the production of food from an average acre of suburban London was worth the same as that from an acre of above-average farmland. This was a comparison of cash values; the weight of food from the suburban land was half that from the farmland, but it was valued at retail prices, whereas the farm produce was valued at farm gate prices.

Nevertheless this was a remarkable finding, especially as only 14% of the housing area was used for growing fruit and vegetables, the rest being taken up by houses, lawns, flowers, paths and drives. This means that the gardens were out-producing the farms three times, in terms of weight of food per area of land actually used for food production. If a larger proportion of the area had been used for food the suburbs could easily have out-yielded the farms overall. Gardening is inherently more productive than farming because of the greater amount of attention that can be given to smaller areas."

– Patrick Whitefield (1996),
How to Make a Forest Garden,
Permanent Publications

TEN TREES WE CAN EXPECT TO SEE MORE OF IN A POST-PEAK AGRICULTURE

by Martin Crawford, Agroforestry Research Trust

Below are ten trees/shrubs with great agronomic potential in future decades in Britain. They are, in no particular order:

Planting an almond tree to create urban food security.

Sweet Chestnut

As preservatives are gradually banned for using on fence posts, we need a sustainable source of naturally durable posts, and coppiced sweet chestnut is the ideal candidate. Every farm should have some! Sweet chestnut also has fantastic potential as an economic nut crop, with the right varieties cropping very well in England already and likely to do even better as time goes by. In our trials in Devon some varieties are yielding 3 to 4 tonnes per hectare. Sweet chestnuts are nutritionally comparable to rice and can be considered a tree 'grain' crop – in the past whole societies have depended on chestnuts as their carbohydrate staple; perhaps they will in the future too.

Apple

Growers are already starting to re-graft to more southerly varieties and this is bound to continue. As the climate warms, regions will have to accept that their 'local' varieties may not be the best for that location any more. In Britain we have already had a climate shift of over 100 miles southwards over the past 30-40 years so that some French varieties are now rather well adapted to southern English conditions.

Bamboo

No, it's not a tree or shrub, but such a useful woody perennial that I had to include it. In Asia, of course, it is used for almost everything. Easy to grow, and very good culinary bamboo shoots are easy to produce. Bamboo canes from China are of course very cheap at the moment, but how long will that last?

Plum

The only other fruit mentioned here, so productive and easy to grow that I had to include it. Plums, like apples, are not native, and most varieties will do better with increasing temperatures.

Downy oak

Widely used in French forestry but virtually unknown here, this oak is likely to become a major forestry tree in the UK as temperatures warm. The timber is of excellent quality. Our native oaks are likely to become drought-stressed in some places and replaced by downy oak.

Walnut

The French can foresee the supply of cheap tropical timber drying up, and are already planting masses of hybrid (American – European) walnut for fast high-quality timber production – we should be doing the same. The right nut-bearing varieties can also be an economic crop in most parts of the country.

Alder

Alders are some of the best nitrogen-fixing trees for our climate. They are also a fantastic windbreak tree, and windbreaks are something that all farmers should be thinking about, because we are going to get stronger storms in future. Italian or red alder grow a metre a year or more and fix substantial amounts of nitrogen, much of which gets released into the adjoining land to fertilise other crops.

Pine

As well as producing good quality timber, pines are an important hydrocarbon source via tapping. Most of the world supply of turpentine still originates from tapped pine trees, and some important products which are currently made from oil could be made from pine resin instead on a sustainable basis. Maritime pine is likely to become an important forestry tree as the climate warms. And some pines, such as stone pine, supply excellent edible pine nuts too.

Willow

Particularly good in easterly windbreaks because they come into leaf early. Short rotation coppice (SRC) also has good potential on a farm scale for heating, though its use in generating energy via SRC power stations is dubious.

Lime

One of our most useful native trees! Coppices very well, one excellent use for coppiced logs is for growing mushrooms, e.g. shiitake, on a sustainable and low-energy basis. Lime trees also have very good edible young leaves – indeed, it is one of our most used salad plants.

Martin Crawford is the Director of the Agroforestry Research Trust in Dartington, near Totnes in Devon. He is the author of many publications on the subject and of the quarterly journal *Agroforestry News*. www.agroforestry.co.uk. He is writing a new book on agroforestry, which will be published by Green Books in 2009.

14th September 2014

LETITIA LLOYD VOTED QUEEN OF CELEBRITY LOVE ALLOTMENT

VOLUPTUOUS CHANTEUSE LETITIA LLOYD was last night declared winner of ITV's ratings-smashing show, *Celebrity Love Allotment*. She won 72% of the vote, knocking out the other finalist, rapper Stig Fresh.

The show began four weeks ago, with 12 celebrities, 6 men and 6 women, moving onto the allotment at Crouch End, London. None of the celebrities had ever gardened before, and to win the show had to demonstrate their aptitude with a range of produce. Initial favourite, soap star Trixie Bishop, came a cropper on her Mizuna greens, and celebrity chef Bob Lard was evicted from the show in disgrace after slug pellets were discovered in his trailer.

An emotional Letitia meets the Press on emerging from the Love Allotment at the end of a trying month.

The show also had its highlights on the love front. **Letitia Lloyd** declared herself heart-broken when her beau (hunky children's TV presenter Nathaniel Ackroyd), was voted off in week three. Viewing figures peaked when Pixie Hargeaves and footballer Dwayne Adams were spotted intertwined behind the runner beans, the 'did they or didn't they' debate keeping the tabloids frothing for weeks. Pixie has since told *The Sun* they were merely practising safe bending techniques for double digging.

The show has triggered an unprecedented degree of interest in food production, with herb-filled window boxes now being the must-have item on souped-up cars in Essex. A new show, *Pimp my Patio,* is now in production. Basking in the glory of winning *Celebrity Love Allotment*, Letitia told reporters as she left the allotment, "I can think of nothing more enjoyable than gardening and growing food . . . having short nails is so liberating! I plan to open a designer salad smallholding in the West End now – *'Letitia's Lovely Lettuces'!* "

blocks,[6] as well as local timbers. This in turn has enabled the creation of small-scale industries to process and produce these materials, often based on the farms.

Others now focus on growing organic mushrooms for both culinary[7] and medicinal[8] uses. Some specialise in growing hemp for fabrics,[9] or producing wood pellets[10] or biofuels such as biodiesel or ethanol for the local market. For some farms, the installation of a methane digester means that they are able to supply heat and power to the neighbouring community.[11] This newly found diversity of enterprise, alongside food production, has led to a much regenerated local economy, with the major part of each community's wealth being cycled locally rather than being leached out into the wider economy.

Over the last 25 years, food and farming in the UK has returned to being regarded as central to the security of the country. Food security is now seen as not exclusively an issue for developing nations. As the rising price of fuel and demand for land for the short-lived biofuels industry began to inflate the price of food around 2011, we found, for the first time in 50 years, that it was cheaper to eat local organic food. At the same time our diets, by necessity, became more seasonal and less meat-based.

Urban agriculture is now a priority for urban planners and communities.[12] Cities have been redesigned as productive places.[13] The city of London now produces 60% of its fresh vegetables and 30% of its fruit in and around the city – and Bristol is now pushing 80%. A massive programme of productive tree planting has brought fruit and nut trees into every park and school grounds. Urban market gardeners began to colonise land around the edge of the cities, producing a diversity of fresh produce for local markets with extremely low food-miles (leading to the use of the term 'food feet'). The keeping of small livestock, particularly chickens, has become the nation's favourite pastime. What were large parks now feature a diversity of allotments, market gardens and horticultural training centres.

Back garden and allotment food production was already a very popular leisure activity in 2008, but in 2012 the Government legislated to make gardening a key aspect of their carbon reduction and health promotion strategies. Now local varieties of fruit and vegetables are highly treasured, and the teaching of intensive organic gardening techniques is a core part of the National Curriculum, as part of the nation's Food Security programme.

Medicine and health

Today our idea of health – how to create it and maintain it – has changed markedly from that of twenty years ago. The Health Service had to rethink itself as the oil price made many of its practices and approaches unaffordable, and it faced the very real threat of collapsing completely. The closure of local hospitals in favour of centralised ones – so rampant twenty years ago – has been reversed, and local healthcare centres are now not just about treating illness but promoting health in many diverse ways. They have forged partnerships with local schools, promoting food growing and familiarising young people with the whole food cycle from seed to salad. The wellbeing of the individual is seen as inseparable from the health of the community. Human biology is now a compulsory school subject, and has expanded to include nutrition and basic herbalism.

About half of the medicines prescribed by doctors are now locally sourced, with local farmers growing certain key medicinal plants which are processed in local laboratories. Local chemists also now make over 50% of the medicines they sell on the premises. Doctors are able to prescribe a range of complementary treatments, as well as involvement in local community gardens, and access to affordable good food. The growth in access to meaningful work, the rebuilding of social cohesion and an emerging common sense of purpose, has resulted in fewer stress-related illnesses and cases of depression. Conventional and complementary practitioners are seen very much as two sides of the same coin, and the concept of promoting health rather than just treating disease has led to a range of innovative measures.

As a result of people's moving away from being sedentary consumers to becoming more physically active producer/consumers, there has been an increase in musculo-skeletal problems. Doctors are now able to issue prescriptions for, for example, Alexander technique sessions. It has become more commonplace, as in China, to see free Tai Chi sessions in local parks in the morning. Technology has also enabled certain tests and observations to take place online in the patient's own home, what is known as 'tele-medicine'.

Education

Education in 2008 was woefully inadequate, given the scale of the Transition to come. It became clear around 2010 that young people leaving school were unprepared for the more practical demands that the emerging powered-down world made of them; their school years had left them unable to build, cook, mend, garden or repair, and the Government declared that youth was in crisis and education needed fundamental reform. A new curriculum was approved in 2012 which re-emphasised voca-

"The coming scarcity of fossil fuels, on top of inflationary costs in medicine (the prices of oil and natural gas are approximately four times what they were in 1999 and rising) and the expenses of treating Baby Boomers (a cohort twice the size of its predecessor), could overwhelm a medical system already in crisis. We can avoid collapse, however, by reducing medicine's present consumption of energy and creating a health-care system that reflects our actual relationship to resources. Ironically, peak oil can be a catalyst for creating a health-care system that is cost-effective, ecologically sustainable, and congruent with a democratic social ethos."

– Daniel Bednarz (2007), 'Medicine After Oil', Orion Magazine

Bristol Evening Post 15th December 2013

GCSE in Sustainable Living Praised

THE HEAD OF OFSTED, Bob Sprout, yesterday praised the results from the first year's participants in the GCSE in Sustainable Living. Local school Henbury Comprehensive produced 70% A grades. Much of this success has been put down to the course's practical element, which included the creation of a community forest garden and the building of a cob/straw-bale classroom.

Principal Michael Curtis said that Henbury (which is set to become the first 'Powerdown School' in 2013), brought expertise into the school that had huge beneficial impacts for all the students. "It gives me great pleasure to know that Henbury School now turns out the finest cobbers, mulchers and daubers in the West Country," he said.

Mr Sprout praised Henbury as an exemplar school. "Many other schools now run the GCSE", he said, "but very few have achieved the level of engagement Henbury has." The school, which is now a net producer of energy, sits in a landscape of productive trees, gardens and ponds. Its aim of producing 30% of the town's fresh produce by 2015 is well on track for being realised.

tional education firmly rooted on foundations of sustainability and resilience-building. From primary school level upwards, gardening, cooking and woodwork skills have become a core part of the programme for the first time since the 1950s. School grounds have been transformed into intensive gardens, with many students running their own enterprises.

By secondary school age, students now learn construction, as well as creating, installing and maintaining renewable energy systems, alongside social skills like conflict resolution and community leadership. For adults, Colleges of the Great Reskilling are now central to most towns, offering a variety of courses in a wide range of practical sustainability skills for the public as well as retraining for professionals.

The number of smaller local schools around the country began to grow around 2015, as the price of liquid fuels made it unfeasible for children to travel long distances to school. By 2018 many of the larger comprehensive schools and universities were no longer able to attract their intakes from large areas and had to rethink how they used their facilities. With unused space on their hands they diversified, and are now also home to incubator units for new businesses, with skilled craftspeople having their workshops and offering apprenticeships onsite. Those schools which have become farms or intensive market gardens also now feature a diversity of value-adding enterprises. Schools are now vibrant, productive, bustling places, firmly rooted in, and key contributors to, the local economy.

Economy

The way the economy works, and the way we think of money, has changed significantly since 2008. The globalised economic model began to run into significant difficulties sometime

Lewes Today   February 15th 2012

Lawyer's Local Lewes Lolly Landmark

By Dave Schmink, Finance Editor

Bob Charles celebrating in the Lewes Arms yesterday.

BOB CHARLES YESTERDAY completed his year's attempt to be the first person to survive for a year without using any Pounds sterling, relying instead purely on Lewes's two complementary currencies, the Lewes Pound and the Time Bank. "I can't say it was the most lavish year I have ever experienced", said former lawyer Bob, 43, "but it was exhilarating and I have made so many new friends through doing this."

Bob started his "Local Lolly Challenge" last April, in order to raise (real!) money for the first local community Land Trust initiative, which is aiming to purchase a 10-acre site on the edge of the town to transform it into an eco-village development and community farm.

Lewes Mayor, Marcella DuPont, congratulated Mr Charles. "Although Bob made a lot of sacrifices over the year, he has shown how far the economy of this town has come since 2007. He has also raised £15,000 'normal' pounds for a very worthy cause." Mr Charles celebrated the end of the Challenge last night in the Lewes Arms with a pint, which he paid for, naturally, with a crisp new Lewes Pound.

around 2010 when world oil production peaked. A period of sustained recession followed: a difficult transition, as our over-reliance on foreign investment and perilous levels of consumer debt became apparent. Parallel to this recession was the vigorous emergence of more localised economies. With national currency in short supply and pension schemes in tatters, towns and cities were forced to develop their own economic systems. New forms of trade are now commonplace, with systems such as LETS and Time Banks flourishing.

Towns and cities, as they did historically in times of hardship, now produce their own printed currencies, only usable within the town. Local investment models have been developed, whereby people invest their money in ways that support the economic regeneration of the community. As the focus becomes increasingly local, people now find the percentage of their daily transactions conducted in national currency continuing to fall. Money is now more answerable to the communities it serves. These local currencies may be backed by the national currency, but increasingly they are backed by locally produced energy or food production.

Each town and city now has its own printed currency used by all local businesses and proudly bearing the heads of prominent local historical figures. As part of national government policies to strengthen local economies, government grants and funding for the community are invested in the local currency and local authorities also accept part payment of Council Tax in local currencies. Shops pay part of their business rates and their local suppliers in them.

As globalised business models have begun to unravel, local entrepreneurs have stepped in to fill the gaps. In the 1930s nearly all businesses were owned by local people; a hundred years later this is true once again. The myth that a strong economy can only emerge if it is based on inward investment is now seen as an oddly warped argument from the Age of Cheap Oil. For communities dependent on globalised businesses, the transition was particularly difficult, but led to a firm commitment to building stable local economies.

Devizes Gazette & Herald

June 10th 2013

HEMP HOODIE HELL

"Can we have our hoodies back please?" asks the Home Secretary's daughter

by Marcella Pruwitt

SANDRA MILTON, co-founder of Devizes eco-fabric company Devizes Dancewear, last night laid down the law to her father, Home Secretary Augustus Milton, in asking for the hemp hoodies seized by police to be returned to their warehouse in the town. Her call followed the seizure by police three weeks ago of 500 hemp hoodies in response to the national hoodie ban imposed in November 2011.

"This is an outrageous attack on people's freedom to wear what they like", she told the Devizes Gazette & Herald. "We have pioneered local clothing manufacture and although we have been very successful this will hit us hard." As well as the hoodies, Devizes Dancewear also makes jeans, T-shirts and jackets, all with local wool and hemp, and buttons are made from local wood. The company now employ twenty people in their workshop on the town's industrial estate. It was these premises that were raided by police last week.

The Home Secretary declined to comment, saying it was a family matter.

Transport

Private car ownership is now no longer the norm. Indeed, other than in some very rural areas, given the extent of the public transport system and the reprioritisation of urban streets to favour cyclists, pedestrians, trams and buses, private car ownership is seen as positively anti-social. The idea that one could live in a rural area and live an urban lifestyle has become a thing of the past. Rural communities have re-organised themselves around the re-creation of local employment, production and community. This has inevitably meant that over twenty or so years the population has changed as those seeking a more active, productive, rural lifestyle have moved out of the cities, while those seeking the enhanced sociability of urban living have headed in the opposite direction. Car clubs are a lot more common, allowing people access to cars without needing to own them; they also mean that cars are better used.

Cheap air flights are looked back on with nostalgia. The inability to travel long distances has had the added advantage that people are more connected to their immediate area, more intimately acquainted with its nooks and crannies. Back in 2007, local people were more familiar with Paris than Exeter, with Delhi than Manchester! Sharp rises in fuel prices and the Government's decision in 2009 to tax aviation fuel sent many of the budget airline companies out of business. Although air travel and the private car were the transport sector's losses, commercial sail-power returned with a vengeance, and other winners included tram and bicycle manufacturers.

Part of the process of relocalisation has been a slowing down from the frenetic pace that typified life in 2008. This has reduced the need to dash off somewhere exotic to 'relax'. People nowadays are more drawn to long summer days on their allotments, sleeping in their summer-houses, taking long cycle rides and familiarising themselves with the ecology and history of their bioregion. Indeed, the transformation of our towns and cities from large, bland places with a few 'entertainment' venues, to diverse places with gardens, ponds, artworks, more opportunities for meeting and working with people, and generally more to see and do, has given people less reason to travel to be entertained.

In 2012, the advent of 'peak cars' (closely following 'peak oil') meant that demand for car parking spaces began to fall, leading to councils looking for different uses for their large expanses of underused tarmac (for whose upkeep they were responsible). Many of these areas were handed back to community control, and became community market gardens and centres for Great Reskilling training. Public transport is now exceptionally well thought out and integrated. Many of the small branch rail lines shut down by Beeching in the 1960s were re-opened, to the great benefit of both the communities and local farmers who can now use them to send produce to local markets. Urban streets now prioritise pedestrians and cyclists, cars having been designed out of many public spaces.

Energy

The UK has reached, thanks to an extraordinary crash programme initiated in 2010, a point of near self-reliance in energy. This was achieved through a 50% reduction in energy consumption and a massive scaling-up of renewables to provide the remaining 50% of energy demand.[14] This was brought about partly by the introduction in 2010 of carbon rationing, based on the model of Tradeable Energy Quotas (TEQs) developed by David Fleming,[15] which allocated

"When I see an adult on a bicycle, I have hope for the human race."

– H.G. Wells

"The choice is clear: if a minority of powerful nations continues to favour an economic system underpinned by centralised technologies and vulnerable supply lines, they will need to protect it at enormous expense and risk to our civil liberties. On the other hand, if we shift to a decentralised world economy based on equitable and efficient use of renewable energy sources, and relocalised supply systems, we will have communities that cannot be easily threatened and, most importantly, which threaten no one else."

– Paul Allen, Director, Centre for Alternative Technology

November 18th 2011

GREEN LIGHT FOR SOLAR ROOFS!

By Nigel Slattery

SOUTH HAMS DISTRICT COUNCIL VOTED YESTERDAY to relax planning controls on the installation of solar panels on roofs in conservation areas in central Totnes. Over 50 applications for solar panels were refused in 2010 and 2011, all on the grounds of being 'inappropriate' for a conservation area.

Announcing the policy U-turn, Cllr Billy Teal said: "The international energy situation is such that we must question even long-cherished beliefs. What is the use of a conservation area if one can't enjoy it because all the lights are out? I think we will learn to love these panels, I'm sure that after a while we won't even notice them."

Susan Simons, who lives in South Street and has spent years battling for permission to put solar panels on her 17th-century house, told the *Totnes Times:* "Finally, common sense has prevailed. It's better for my blood pressure not to think of the carbon that could have been saved if this change had been made five years ago. My application will be on their desk first thing Monday morning."

to each citizen and business a carbon allowance which was gradually reduced each year, and managed electronically with a swipe card, used with every purchase of energy or fuel. Since their introduction, TEQs have rapidly become a fact of life, with some people now actually making part of their income by living simply and trading their surplus quotas.

A nationwide crash programme of domestic energy efficiency and retrofit begun in 2009 has brought down domestic consumption by 60%. Part of the success of this was the mainstreaming of energy efficiency. While domestic solar panels and wind turbines became seen as 'must haves' around 2010, as prices came down and there was increased grant aid, the much less glamorous work of retrofits still needed a push.

This was, in part, achieved by engaging local artists, who reconceived the installation of insulation and other energy-saving devices as a settlement-wide art installation, akin to the artist Christo[16] wrapping buildings and islands. The remaining energy demand has been made up from a mixture of wind (as much as half of it), including a big programme of offshore wind projects, as well as biomass-fuelled Combined Heat and Power systems and tidal power. Many towns have helped to reduce their demand on the grid by creating localised energy mini-grids, often owned and managed by locally owned energy companies using the ESCO model,[17] an approach first tried many years ago in Woking, Surrey.[18] These bodies put in place renewable energy infrastructure which is owned and supported by the community.

These mini-grids are powered by whatever has been identified as the most locally appropriate energy sources, be it tidal in coastal areas, biomass in the Forest of Dean, or wind in the Scottish Highlands – usually a mixture of a range of sources. They are connected to the National Grid in order to exchange surplus or obtain backup when necessary, but communities generating their own power in this way have developed an important tool for strengthening their local economies, enabling the money from its generation to be retained in the local economy.

It is standard practice now that many homes, especially new-build ones, are net energy exporters, thanks to generous subsidies for solar power (passive solar and photo-voltaics). The surplus energy generated is fed into local mini-grids where they exist, or into the National Grid. Every home is fitted with a smart meter, which allows the occupants to see at a glance how much energy is being used in the house at that moment.[19] Energy companies also use tariffs in imaginative ways, charging less for electricity at certain times so as to encourage less peak demand at other times.

People look back over the last twenty years with a sense of enormous achievement. What looked like an insurmountable challenge in 2008 has been tackled with a united effort and with great imagination. People look back to the wastefulness of twenty years ago with astonishment and a certain amount of distaste. The new energy economy is leaner, but people now appreciate that one's degree of personal happiness does not directly correlate to the size of one's energy consumption.

Housing

The nation's housing stock, which although it looks to all intents and purposes much like it did in 2008, is today far more energy-efficient. In 2014, the model of the Local PassivHaus became the standard for all new domestic construction across the UK; based on pioneering research by Rob McLeod in 2007 which combined the technological advances of the European PassivHaus concept with the use of locally sourced biomaterials.[20] This model allowed the construction of homes which require no space-heating at all, drawing all their heating requirements from solar gain and the occupants' body heat. In the local version, in excess of 80% of the materials used are locally sourced.

This in turn has led to an explosion of local industries producing clay plasters, natural insu-

---

Lincolnshire Echo   **November 14th 2014**

TOP AWARD FOR LINCOLN HEMP BUILDING FIRM

By Nigel Tavish

LINCOLN COMPANY 'HEMPIRE BUILDING' was last night awarded a prestigious award for innovation in Building at the Department of Zero Energy Building (DZEC)'s annual awards ceremony in London. The company, brainchild of Lincoln-born Evan Field and Michael Spicer, has developed a building system which is now being used in the retrofit of thousands of homes in the area, especially in and around Lincoln itself.

The company, based on the Industrial Estate in the town, has pioneered the use of hemp-based building materials, with much of the hemp being grown on local farms. The award was for their product, 'HempSlab', a panel board with excellent insulating properties, made from hemp, clay and lime, which can be used to internally insulate buildings, while also improving the internal feel of the room.

The DZEC award recognised the product's dedication to local markets, low embodied energy and thermal efficiency. Rather than expand their Lincoln manufacturing plant, Field and Spicer aim to start franchises around the country as part of their dedication to supporting local markets and reducing carbon emissions from transportation.

---

lation and cob/hemp blocks. The breathable construction and the materials used in the Local PassivHaus has led to buildings that are very healthy to live in, with very low embodied energy; they also contain significant amounts of stored carbon and contribute very few polluting 'building miles'.[21] All new buildings are designed to be autonomous and off-the-grid for

---

"When we build, let us think that we build for ever. Let it not be for present delight, nor for present use alone, let it be such work as our descendants will thank us for. And let us think, as we lay stone on stone, that a time is to come when those stones will be held sacred, because our hands have touched them, and that men will say as they look upon the labour and wrought substance of them, 'see, this our fathers did for us.'

For indeed, the greatest glory of a building is not in its stones, nor in its gold, its glory is in its age, and in that deep sense of voicefulness, or stern watching, of mysterious sympathy, nay, even of approval or condemnation which we feel in walls that have long been washed by the passing waves of humanity. It is in that golden stain of time that we are to look for the real light and colour and preciousness of architecture."

– John Ruskin (reprinted 1961), *The Seven Lamps of Architecture*, Noonday Press

"There is some of the fitness in man's building his own house that there is in a bird's building its own nest. Who knows but if men constructed their dwellings with their own hands, and provided food for themselves and their families simply and honestly enough, the poetic faculty would be universally developed, as birds universally sing when so engaged?"

– Henry David Thoreau (reprinted 1997), *Walden*, Beacon Press

"Most of the buildings most of us live and work in are soulless, anti-ecological, and ugly. We close our senses when we are in them. But there's another kind of architecture, one that feeds the soul and spirit, that helps us feel good, that elevates our daily lives. The old Dominion-over-Nature days are past; we need an 'ecological architecture' that re-establishes us in our place in Nature, where we are constantly reminded of the glory of the world around us. In creating natural buildings, you will create ecstatic spaces, a place where your spirit can soar."

– Ianto Evans (2005), *The Hand-Sculpted House: a practical and philosophical guide to building a cob cottage*, Chelsea Green

water and sewage needs, as well as producing more energy than they consume collectively (as with a row of terraced houses for example) or, for stand-alone buildings, independently.[22]

New models for inhabiting larger buildings, and new living arrangements – such as co-housing, where people have private units but some shared facilities – have become far more common. While the retreat in house prices of 2009 resulted in hard times for many, they also made home ownership feasible for young people again, as did falls in the rates of second home ownership.

The average footprint of new-build homes has fallen, and one of the great arts for architects is now the efficient design of the small home. Years ago, one's sense of social worth was based on the size of one's house; now it is based on its compactness and efficient design. In rural areas, in response to the demands of adjusting to the needs of a much expanded agricultural workforce, clusters of small, low-impact buildings, built from local materials, have been created on farms. Agricultural ties have been used to keep these from becoming privately owned,[23] based on the '15 Criteria for Sustainable Development in the Countryside', developed by the rural planning reform organisation 'Chapter 7'.[24]

In 2011, the Government initiated the concept of the Great Reskilling in the training of construction industry workers. Added to the skills taught were the skilled use of hemp and lime, cob blocks, and so on – a much broader set of skills than had previously been taught. A trip to the local builder's merchants today presents the builder with a range of materials vastly different from those of 2007: bagged clay plasters, straw and clay, reed and clay boards, hemp or cob blocks, lime or clay renders, laths, locally made natural paints, pigments from local clays, and a wide range of locally grown and sawn timbers, as well as underfloor insu-

lating pellets made from expanded recycled glass. Recycling has changed from 2007, when it involved long distance transportation for centralised industrial processing, to being primarily focused on local, low technology reprocessing, and many new innovative building materials are now made from the low-tech recycling of plastics, paper, fabrics and glass.

North-West Evening Mail   **March 21st 2021**

THE LAST DROP
"I felt like a chip" says biodiesel producer.

MOUNTBARROW SERVICE STATION NEAR ULVERSTON yesterday pumped its last litre of biodiesel. The golden liquid that promised so much in 2009 has struggled to establish itself and the UK biofuels industry recently admitted that, "Biodiesel, unfortunately, has no future."

It had all started so promisingly. Government subsidies led to biodiesel filling stations opening around the country, and by 2016, 30% of UK arable land was under biodiesel production. However, the oil shocks of that same year led to the Government's National Food Security Programme, which saw biofuels edged out as food production took priority.

Paul Tuckett, who farms near Swathmoor talks ruefully of the 'Golden Age' of biodiesel. "We were producing 40,000 gallons at one point," he said, "but now we just have a small patch for our own tractors and for the local fire brigade. Other than that we are now a mixed walnut, chicken, pig, myrtle berry, Szechuan pepper, olive and carp farm." Asked which he preferred, he said there was little he missed about being a monocultural biodiesel grower. "It got everywhere," he said, "I felt like a chip for fifteen years."

Mike Johnson of Mountbarrow's said that he would miss the fuel. "I don't know what we'll sell at the pump now," he said. "With oil at $250 a barrel, selling petrol is now such a limited market that only the aristocracy can afford it. This isn't enough to sustain us in a business. With the increase in horse carriage use we are thinking of converting to being a blacksmith and hay-selling business. I suppose we must move with the times, but it isn't always easy."

HELLO!

March 27th 2029

AS THEY MOVE INTO THEIR FABULOUS NEW NATURAL HOME

DAVID & VICTORIA BECKHAM

TELL US WHY HAPPINESS IS A WARM COB BENCH

David Beckham last night carried his wife Victoria over the threshold of the home they have built for their retirement, and told *Hello!* Magazine that they were "delighted, really thrilled. It's the most gorgeous house we've ever lived in. It's so us."

Two years ago, David, 54 and Victoria, 55, decided to retire early and to focus on their passion, growing heirloom vegetable varieties. Last week, when this reporter visited, the back of the house was full of potted herbs, cuttings and trays of seedlings brought from their previous garden. David has also set himself the aim, before he reaches 60, of mastering the art of hot composting.

The couple's house takes the new fashion for small, compact and well-designed spaces to a new level. Built with 3-feet thick cob and hemp walls, the thatched house is, like Posh Spice in her heyday at the turn of the century, curvaceous and alluring, the couple deciding at an early stage that they didn't want the house to contain any corners. "We wanted a house which was dead cheap to heat, used local materials and stuff, was off-the-grid, and where walking into each room felt like walking into a hug. Corners are all very well on a football pitch," David jokes, "but since my retirement I certainly don't want them in my house!"

Victoria excitedly adds, "We've got solar panels, a masonry stove thing for heating, a really cool composting toilet and a fridge which works without electricity, just by drawing cool air through the ground. It's all really clever."

With their sons now grown up, David and Victoria decided three years ago that they only needed enough space for the two of them. "We wanted a house that fitted us like a glove," David tells *Hello!* Magazine. "As we designed it, using clay models, our architect kept pushing us to think smaller, smaller, all the time. Some of his ideas were amazing, the sleeping loft over the kitchen, the drawers in the stairs, the niches and alcoves, they were all his input."

The house puts the couple at the forefront of the fashion for small and well designed homes. Katie and Tom Cruise recently tried six months in a yurt, but found the winter months hard going and in the end felt drawn back to their three bedroom straw-bale house. Chart-topping singer Letitia Lloyd is experimenting with Earthships in Essex, and Charlotte Church's roundhouse in Wales is a highly individual celebration of hemp construction.

Back at what David and Victoria, still clearly deeply in love after 31 years together, call their "Love Shack", David is sitting on the cob bench he helped to create. "Look at it", he says, "it's so sculptural. After a hard day mulching the garden, I love sitting on this bench. Look,

David's gardening books adorn this gorgeous cob niche in the Beckham residence.

I've my gardening books in this niche here, and when we fire up the stove, the air runs through and warms the bench. What more could I want?"

What more indeed. David and Victoria are, as ever, fashion trailblazers, darlings of a post-petroleum age. Snuggled up together on their heated bench with a bowl of fresh mixed salad from the garden, David muses: "When I look back at photos of us twenty years ago, given all that has happened since, I have to wonder, as I sit here on my warm cob bench, 'What were we thinking?'"

"Traditional materials, using vernacular technologies are generally appropriate to local conditions, being drawn, for the most part, from available resources. . . . Their use also has important economic implications, being obtained with the minimum of transportation costs and, frequently worked by hand, involve relatively little use of high-energy-consuming fuels. Vernacular architecture is therefore ecologically sensitive and with regional resources being carefully nurtured, is, as it has always been, economically and environmentally sustainable."

– Paul Oliver (2003),
Dwellings: the Vernacular House World Wide, Phaidon Press

Tools for Transition No. 6: Using visioning as a tool in peak oil education in schools – 'Transition Tales'

How might one introduce peak oil and the concept of Transition to young people? What follows is a collection of tools that were developed for working with students at King Edward VI Community College in Totnes, as part of the Transition Tales project. These tools are still being developed but I hope you will find them useful.

Session 1:
The "What is peak oil?" session

Produce a large bag containing all kinds of household objects -- trainer shoes, hair gel, inner tubes, spoons, and so on – and ask the students what all these things have in common. Let the suggestions fly for a while until you have unpacked the whole bag, and then, if no-one has got it, tell them that they are all made of oil. Then whip out a litre bottle of oil, and make the point that a couple of tablespoons of it contain more energy than they could exert in a day, and that their lives require the equivalent of 50 people in their back garden pedalling frantically on bikes day and night. Then split them into 5 groups, each group on a separate table with a different exercise on it. Move them round to another table every 15 minutes.

Table 1: The journey

Where does oil come from and what does peak oil mean in practice? What kind of state is the supply we depend on in? For this exercise, you'll need a computer online, logged onto David Strahan's Oil Depletion Map (www.lastoilshock.com/map.html), and to have given each group a sheet (see Appendix 1, p.214) which asks them six questions they could only find out the answers to by roving around the map. This gives a good background and gets them thinking about what peak oil means and in particular how vulnerable the UK is becoming.

Table 2: War and oil

On this table, divide the group into two teams, and invite them to debate the motion, "the war in Iraq was caused by oil". Cards with various reasons, such as 'WMD', 'oil', 'democracy' and so on can be arranged on the table, and the students invited to rank them in order of importance as they see them.

Table 3: What you can do with oil

On the table with the contents of your bag of oil-made items place a card reading: "Oil can be

used to make a vast array of plastics, glues, paints, varnishes, medicines and other materials, most of which we take for granted in our lives. On this table are many of the things that we looked at earlier. Your task for the next ten minutes is to find at least one thing in this room that is not made using oil. If you can't do that, try to think of something at home."

You will need to think about all the elements and components of the objects you are thinking of."

Peculiarly, the group we did this with came up with three things, which were: "ourselves, hair, and the water in that jug".

Table 4: How is oil used In Totnes?

On the table place a pile of copies of the local newspaper and some red marker pens, as well as some photocopies of copies of the same paper from the 1930s (some rummaging around in the local library or museum should provide these). In pairs, ask the students to go through the papers and circle any article which involves the use of oil, including adverts. Using the 1940s cuttings offers an insight back to a time when the pressure was to use less, not more, and the contrast between the two can be quite illuminating.

Table 5: Life beyond oil

The card on this table reads "whether we see life beyond oil as a challenge or an opportunity will be

down to us. Do we choose to see the move away from our dependence on it as a disaster, or as the opportunity to build something better?" Prepare sheets of flip-chart paper with two columns, "What would you look forward to?" and "What would you miss?", and invite the students to brainstorm things for each column.

Session 2: Transition Tales

The next session turns to storytelling. It begins with a series of exercises designed to free up the students' creative expression and imagination.[25]

1. Stick stories

Divide students into groups of five, and each group is given a stick, just a normal stick you might find under any tree. They are given one minute to talk about the stick, without hesitation, deviation or repetition, passing to the next person after the minute. They start by just talking about the stick, but after a few goes, amazing tales began to emerge!

2. Tell the biggest lie!

Next, invite the students to choose any object from around the classroom, and then to tell outrageous lies about it and what it is for. The outrageous and ludicrous become mixed with the mundane and occasionally outrageous.

3. What are you doing?

Get the students into pairs, one then has to mime an activity and the other has to ask what they are doing. The student who is miming has to reply with another entirely different activity that the questioner then has to mime. This goes backwards and forwards for a while.

4. "In the city of Rome there was a street . . "

This activity is like the old game "I went to the market and I bought . . ." where you say one thing

Tools for Transition No. 6 (continued)

you got at the market, the next person has to say yours and then add their own, and so on around the group, getting longer and longer each time. The rule of this version is that you add things that are observations, descriptions, rather than nouns, so that rather than telling a story, you are building up a descriptive picture of a scene.

5. The black box

Sitting in pairs, students are asked to imagine a black box between them, and to begin with, they go round the group, putting their hands in the box and pulling something out. They are asked not to think about it too much, to have no idea in mind what is in the box when they start to put their hand in, and just see what emerges. The second refinement to this is that you put your hand in, draw something out, and then the other person has to draw a story about the object out of you by asking questions, to which you can reply in the most fantastic ways, so that together you create a story about the object.

The news from 2030

In groups of five, the students are then set the task of devising a TV news item from their powered-down community in 2030. They can do news bulletins, interviews, the weather, or whatever they choose. This is really the proof of the pudding in terms of whether they have really engaged with the previous two sessions. Film their presentations (something about having a

camera makes them focus a lot more), and then edit them into a news bulletin. Remember if you are to use the footage, you will need written parental consent. When we did this in Totnes, one group did a 'Top Gear'-style car review of the latest transport sensation in Totnes, which consisted of hopping on his friend's back and having a piggyback round the room. Another group did the weather ("cloudy and windy, so if you have solar panels, bad luck, you'll have to wait until tomorrow to do your washing, if you've got a wind turbine, you'll have a great day") and one group of girls all dressed up in Edwardian dresses with lace parasols and bonnets and said that they didn't need sun cream any more because they used parasols instead.

To view the final Transition Tales film, visit www.youtube.com/v/9c6ubbq4Hzo&rel=1.

Oxford Times   **15th March 2011**

TAKING THE P★★★!

"We stopped being squeamish and started being practical." By Paul Haig

VISITORS TO OXFORD this year may notice that things have changed a bit since their last visit . . . at least they will when they go to the loo! The city's public toilets and several of the hotels have installed urine-separating toilets. The loos themselves don't look very different from what was there before, indeed the urinals are almost identical. It is what happens after use that they may be wondering about.

The new set-up is the initiative of N • Pee • K, a new Oxford-based initiative. Why would anyone want to collect the liquid doings of the town? N • Pee • K director Imelda Platt explains: "With North Sea natural gas almost entirely depleted, nitrogen fertiliser production is being hit hard. Nitrogen is essential for our farming community, and is becoming unaffordable. The average human produces roughly the same amount of nitrogen in their urine as agriculture requires to grow their food. We decided to stop being squeamish and start being practical."

Each participatory institution has a large sealed tank, reminiscent of the home heating oil tanks some of you may remember from a few years ago, tucked discretely behind the building. The tank is emptied

Imelda Platt proudly displays her company's 'Liquid Gold', harvested from Oxford's pubs and hotels, which has led to N • Pee • K becoming the city's fastest growing business in 2010.

every two weeks, diluted and sold on to local farmers. Platt calls her product 'liquid gold'. "When I approached all these places five years ago and asked to buy the rights to their urine, they thought I was mad," she said. "We paid to install the loos, it was an investment that had paid back within six months. This is a business with a future," she concluded. "We aim to continue expanding. Unlike natural gas, there is no danger of our raw material drying up!" [26]

Chapter 9

Kinsale — a first attempt at community visioning

The Kinsale Energy Descent Action and how it came about

Kinsale (*Cionn tSáile* in Gaelic) is a town 25 miles south of Cork City in County Cork, Ireland. It has a population of around 2,300, and is a popular destination for tourists, both Irish and international, who come for fishing, sailing, golf and the gourmet food for which it is known. In September 2000 I initiated a permaculture course in the town's Further Education College, which already had a strong reputation for art, drama, and multi-media courses. In its first year the course was essentially an expanded permaculture design course, but over the next three years it grew to become a two-year full-time course with modules in permaculture design, organic growing, natural building, sustainable woodland management, conflict resolution, community leadership, starting your own business, nutrition and field ecology. It is one of the first such courses in the world.[1]

The course was, and still is, extremely popular, and attracts students from around the world. The grounds of the college have been transformed, as a result of the inspired hard graft of generations of permaculture students, from being an expanse of lawn to a diverse permaculture landscape, featuring a small coppice woodland, a straw-bale house, a polytunnel and vegetable beds, a forest garden, a small pond, and its crowning glory, the cordwood amphitheatre, built entirely using local materials.

In September 2004, on the first day of term, the students were shown a new film called *The End of Suburbia* (the first time I saw it), and then, later that day, Dr Colin Campbell of the Association for the Study of Peak Oil gave a talk to the students. The combined effect of this 'double whammy' was very powerful for the students. It greatly focused the mind, and came as

End-of-year photo 2005, showing cob/cordwood amphitheatre just completed in time for first performance.

quite a shock to everyone – myself included. 'Post-petroleum stress disorder' manifested itself in diverse ways. Once the mental dust had settled, the concept emerged of a project for the second-year permaculture students, which (on the assumption that Campbell's forecasts would be proved to be true) set out to explore how the town of Kinsale might successfully make the transition to a lower-energy future.

The first thing I did when conceiving the project was to look around for existing examples of this already happening. Astonishingly, we could find nothing at that point that constituted a community response to peak oil, despite scouring the internet. There were some useful tools in James and Lahti's *Natural Step for Communities*,[2] some initial thinking by the Post Carbon Institute, and the two key books that most inspired the process, Richard Heinberg's *Powerdown* and David Holmgren's *Permaculture – principles and pathways beyond sustainability.*

Powerdown takes the reader on a journey exploring humanity's options beyond the peak. Heinberg argues that the desirable outcome lies somewhere between what he calls 'Powerdown' (a national government-led programme of economic contraction and relocalisation) and 'Building Lifeboats' (communities responding by building local infrastructure and self-reliance). While the book contains little in the way of practical suggestions for how this desired outcome might be achieved, it remains a seminal analysis of humanity's options, and the concept of 'Powerdown' has entered the collective consciousness. (The brief I set for this project is given in Appendix 3.)

Having received their project brief, the students divided into pairs and took on different topics. They drew up a list of the relevant people in the town, and I suggested useful reading and websites. We invited these people to come in and speak to the group, and went on a number of visits to permaculture sites, organic farms, and green buildings, to ask for the input and insights of those who have been practically engaged in this kind of work in the area for many years.

On Saturday February 12th 2005 we held an event in Kinsale Town Hall called 'Kinsale in 2021: Towards a Prosperous, Sustainable Future Together'. The event was presented as a 'community think-tank', in order to hear the community's ideas about how energy descent would affect the community and what might be done about it. We sent personal invitations to the movers and shakers in the town, drawn from the sectors identified in the project brief. The event was also open to the public, and we put posters up around the town. From the 60 people invited, about 40 turned up on the day. The event itself was opened by the then Mayor of Kinsale, Mr Charles Henderson, who spoke of the importance of energy as an issue and how it affects all aspects of our lives and our economy. This was followed by a screening of *The End of Suburbia.*

After the film we introduced the concept of Open Space Technology[3] as a tool for facilitating community exploration. People were invited to identify the specific problems and issues that the film raised for them. These were then recorded on large sheets of paper, pinned up on the wall and then collated into subject areas, each of these becoming the basis for a discussion group. The groups covered the following subjects: Food, Rebuilding Communities, Youth Group/Education, Business & Technology, Tourism and Renewable Energy. The groups came up with a wealth of ideas and possibilities that were then fed back to the rest of the participants.

After the event we collated the information that had come in from the day and pairs of students selected different subject areas. I supplied a lot of reading material for background research, and the students did a lot of internet research for useful ideas and examples from around the world. The final result was *Kinsale Energy Descent Action Plan – Version 1, 2005* (KEDAP),[4] which attempted a year-by-year plan for the town. Each section of the report begins with a section called The Present. This attempts to succinctly summarise what is the problem now, in 2005, with regards to the subject in question.

The Present is followed by The Vision, which is written in such a way as to give the reader an idea of how Kinsale could be, if all the recommendations up to that point had been implemented. This is then followed by a list of suggestions and recommendations, in chronological order. These are meant to be ambitious but also achievable, given a good deal of commitment and support. Each section is then rounded off with a collection of resources and internet links. I then edited and designed the KEDAP and 500 copies were printed in time for the Fuelling the Future conference run in Kinsale in June 2005. This conference featured, among others, Richard Heinberg,[5] David Holmgren and Colin Campbell. It was a memorable and impactful two-day conference which explored community responses to peak oil.[6]

At the time I don't really think we had grasped the significance of what we had created. The KEDAP wasn't even formally launched at the conference; it was just made available to the delegates on a table at the back of the marquee. It wasn't mentioned in the local press for some months, and slipped out rather than actually being launched. The copies sold steadily however, and fairly soon the only places it was available were as a download on the Fuelling the Future website and at www.transitionculture.org.

Kinsale 2021
An Energy Descent Action Plan – Version.1. 2005

By Students of
Kinsale Further Education College

Edited by Rob Hopkins

Kinsale Further Education College
Oideachas Tré Eachtra

It has since been downloaded many thousands of times, and has inspired many similar initiatives around the world.

Four lessons from the Kinsale Project

In the months following the publication of the KEDAP, it became clear that we had stumbled across something important and powerful. People from all over the world got in touch to say that it was the missing piece of the puzzle, the thing they had been looking for. As interest in it grew, I started to muse on Transition culture, about what lessons might be discernible from the process that could inform similar processes. The resultant insights have done much to underpin the work subsequently undertaken across the emerging Transition Network.

Lesson One: Avoid 'Them and Us'

It would have been easy to fall into the same trap that so much campaigning and activism falls into of creating a sense us 'them and us'; painting the local council as being the villains of the piece because they had failed, thus far, to begin formulating responses to peak oil, and indeed had done very little that could be called 'green' at all. It could also have resorted to attacking the business community for its un-green ways, but instead it sought to involve them.

Members of the Council and other stalwarts of the community were invited to the Open Space day as well as to the 'Fuelling the Future' conference. Many of them were approached and asked for their views, and the Town Hall was used to host the Open Space event, with the Mayor opening the proceedings. The project was always carried out in a spirit of inclusion and openness. The Practical Sustainability course attracted, from the outset, a wide range of people, many of them from Kinsale itself, who helped spread the concepts around the town.

The more I have been involved in this work and met people working in positions of authority, be they planners, engineers, councillors and even politicians, I have seen that they are ordinary people, often with families, just as bewildered by the turn of events as everybody else. For us to scream "why aren't they doing anything?" does nothing to help.

Lesson Two: Create a sense that something is happening

From its beginning, the Practical Sustainability course developed a reputation in the town for being a place where unusual yet fascinating things were occurring. People often commented to me that they loved the 'buzz' around the town created by it. The various building projects that took place at the college were greeted with great interest, indeed sometimes students would spend the morning cob building or clay plastering, and then head down to the town for some lunch, leading to their being fondly referred to in the town as the 'Mud People'. We also had an annual Open Day where visitors would eat pizza from the clay pizza oven built by the students, salad from the polytunnel, have guided tours and generally soak up the atmosphere.

The amphitheatre project at the College which was completed in May 2005 did a great deal to put the College on the map. Many who attended the performances of 'The Merry Wives of Windsor' put on by the College's drama students, talked about the 'magical' feeling of the space. The College's combination of interesting courses and ground-breaking practical projects did much to make the community well disposed to the Action Plan process when it began. When it came to developing the Plan itself, people were delighted to have input into it, and the reputation that had been built up for work that was exciting and done with good heart went before it.

Creating this sense that something is happening could also be done by staging an uplifting and high-profile conference, as was done in Kinsale with 'Fuelling the Future'. An event such as this, if designed with lots of outreach elements, and also involving local people with relevant skills and knowledge, can really embed the process in the town, bringing in new information and inspiration, while at the same time reinforcing and reaffirming the work that others have already been doing.

Another way of creating such a 'buzz' is through developing a reputation for addressing concerns that are seen as important more widely than peak oil. For instance some of the course's students organised a clean-up of a local nature area which was reported in the local press. Some others designed and planted a permaculture garden for a local hostel. One of my favourite examples of this is the City Repair organisation in Oregon in the US, who run a festival every year called the Village Building Convergence, building cob bus shelters, community gardens and what they call "intersection repair". They create a wonderful sense of innovation and positive things happening.[7]

We need to communicate that this 'something' is profoundly meaningful and transformative,

and has a sense of magic and a spark of wonder to it. Creating this atmosphere is the oil that lubricates the engine of your energy descent process. The more you can create a feeling that something important, positive and dynamic is underway, the less friction and resistance your work will encounter.

Lesson Three: Create a vision of an abundant future

One of the essential ingredients in developing community responses to peak oil is enabling the community to create a vision of how the future could be. We move from working with peak oil, which is about probabilities (how probable is it that it will be horrendous, how probable is oil peak in 2007, and so on) to possibilities. The shift is subtle but illuminating. The Open Space event we ran in Kinsale gave the community permission to dream. It was very powerful to see it happening, people going home excited about how the future could be, and feeling they had met some kindred souls

The opening night of 'The Merry Wives of Windsor, in the cordwood amphitheatre.

KFEC Principal John Thuellier at the official opening of the amphitheatre in May 2005.

with whom they could create a future.

It is important that people can see where they are going and that they like what they see. If we present people with a vision of disaster and social collapse, what incentive do they have to do anything? This is not to say that we should not aim to raise awareness and talk about the issues, but at the same time, simply presenting people with bad news and expecting them to respond by engaging boldly and imaginatively is unrealistic in the extreme. Ken Jones, the Buddhist author, writes of 'changing the climate, rather than winning the argument'.

Lesson Four: Design in flexibility

The Australian permaculture teacher Dave Clark speaks of his experiences implementing permaculture in refugee camps in Macedonia. He was dealing with large numbers of people moving to places with no infrastructure, all of which had to be created. He did amazing work, erecting straw-bale buildings, food gardens, putting in miles of

swales[8] and hundreds of thousands of trees. One thing he said really stayed with me. He spoke of having to work with professional engineers who would design something such as a drainage system, which Dave could see wouldn't work, but which, because the person was a 'professional' could not be questioned. He saw much money wasted through this unchallengeable 'rule' that the professional is always right. He talked about how in his work he always worked from the premise that he was wrong. Designing this approach into the process led to an openness to reassessing at any stage.

I came to think that an Energy Descent Action Plan should be like this. It ought not be cast in stone, rather be a collection of ideas that is reworked and revised regularly. The original idea in Kinsale was for an annual revision, with each new update containing news of what had been achieved since the last one. From the experience of those now driving forward the Kinsale process, this has proved to be too onerous a task on top of actually implementing the plan – which is, after all, the main point. The principle still applies though, that once the first plan is done, it is taken out into the community and 'tested' through a series of specific Open Space events.

The worst-case scenario is what happens with some other 'plans'. They become, like the work of the engineer referred to above, carved in stone, immutable and fixed. "We are working our way through the plan", even though that plan may be long since irrelevant or based on assumptions that have long been superseded by events. By designing this flexibility into the process, we can make it infinitely more powerful, and give the community a far stronger sense of ownership and involvement.

Reflections on the Kinsale process

The KEDAP arose from a programme of community think-tanks, awareness-raising, the work of the students and the inputs of various people in the area with ideas to offer. It is useful to wonder what might we have done differently, and indeed, whether it has made any difference. Firstly, some reflections on how it might have been done differently.

The process was not really as embedded in the community as it could have been – certainly nothing compared to the scale of the subsequent work in Totnes, Lewes, Stroud and the other emerging Transition Initiatives. It was principally my initiative and concept (with the support and inspiration of an assortment of others), and despite many of the students having been involved, when I left at the end of term to move to Totnes, no obvious support structure was in place to enable a continuation of the process. The team who had driven it forward were mostly not from Kinsale, and there was no community group in place to carry it forward.

Had it not been for Louise Rooney and Catherine Dunne (both former students), who formed Transition Design, continuing to drive it along, it could have disappeared without trace. From this I learned that it is essential to root the process in the community. I can however excuse the approach used in Kinsale, in that the Action Plan is, in effect, a suggestion for further action, and once finished was offered to the community to take and develop in whatever way they felt most appropriate. Basically the lesson here is that the process needs an in-built resilience, so that one person can drop out without affecting the project, something that has been better addressed in subsequent Transition Initiatives.

Insufficient awareness-raising

What would have been ideal would have been to have trained students up at an early stage to run peak oil awareness workshops in schools, youth clubs, gardening clubs – to anyone who would listen. The reason we didn't do this in Kinsale was constraints of time. To start with, we didn't know what the process we were developing was going to look like, and were still coming to grips even with what energy descent would mean in practical terms ourselves. Now, in the Totnes process, we run the Skilling Up for Powerdown evening class, to better equip people to be able to do this. This also makes the concept better embedded in the community. The problem with

Students visiting a local organic farm as part of their research for the Kinsale EDAP.

Kinsale was that I didn't live there, so I didn't know all the groups and networks, something such a training course may well have got round.

What's happening now in Kinsale?

Is Kinsale now an Irish coastal utopia, all local food and pushbikes? No. Has the Kinsale Energy Descent Plan been rigorously implemented step-by-step? No. It has however, been the trigger for lots of action and activities. Initially the process was driven along by Louise Rooney supported by Catherine Dunne[9] under the banner Transition Town Kinsale (TTK), and after a while a committee was set up which includes, among others, a member of the Town Council. TTK has monthly public meetings and publishes regular updates in the local community newsletter under the heading 'Transition Times'. They have also:

- run a number of film screenings and talks, as well as a one-day workshop with Brian Weller from Willits, a town in California spearheading innovative strategies for economic relocalisation[10]

- been awarded some grant funding for a community garden project, begun developing a model permaculture garden on a local Council estate

- set up a community composting scheme on another housing estate within the town

- organised a fundraiser for TTK at the Further Education College which generated over 1,000 euros, with food, talks, films,

Kinsale students and local residents working on the pumpkin patch at the Kinsale community garden.

Photo © Nicholas Harvey

and concerts in the amphitheatre

- prepared a brochure about what they are doing (to be distributed to every house in the town), and one member has produced an edited version of the KEDAP, to make it more available and accessible

- run a number of workshops with local schools setting up vegetable gardens and planting fruit bushes.

There are now five subgroups meeting regularly: Food, Energy, Transport, Events, and Waste. When I asked Councillor Isabelle Sutton, a member of TTK, for her thoughts on how the process had gone and whether the EDAP had been a useful thing or not she replied, "It has been a huge help; we wouldn't be where we are now without it. We are always referring to it for the next thing to do."

One hindrance to the Kinsale process, and perhaps the key reason why it hasn't progressed further, is that, in terms of the twelve steps for a

Transition Initiative process set out in Part Three (see page 148), Kinsale started straight away at number twelve, without the preceding eleven. This in effect meant that the process wasn't really grounded in a strong community awareness, didn't have the momentum behind it that an Unleashing generates, didn't emerge from local groups, was only based on one general Open Space Day, and had done no work with the older members of the community. There were, via the Permaculture course at the College, a good few visible, practical manifestations in place, and the Great Reskilling concept was understood and pursued, but both of these were, at that point, quite limited to the College grounds. Because of this, TTK really had to go back to square one, starting the process from scratch with the KEDAP to refer to as a guide, but not as a plan of action endorsed by the community.[11]

However, as we have seen, up to now at least its impacts have been felt far more keenly beyond Kinsale than they have been in the town

Photo © Nicholas Harvey

A Christmas community nut tree planting organised by Transition Town Kinsale. December 2007.

itself.[12] What emerged from the work in Kinsale was an idea, a model, and a sense that, as Richard Heinberg put it in his Foreword to this book, a successful response to peak oil and climate change looks "more like a party than a protest march".

Summing up Part Two

At one of Transition Town Totnes's Open Space Days in late 2006, a woman said to me, "When I think of what TTT is doing I feel so full of hope I could cry." This chapter has argued that hope is one of the key emotions we need to nurture and sustain in order to navigate the troubled waters ahead, and it has been something that environmental campaigns until now have struggled to generate. We saw in this chapter the different ways in which an awareness of peak oil can affect people in the varied manifestations of 'Post-petroleum stress disorder'. We also saw how important the creation of visions of a lower energy future is, and explored some tools, such as storytelling, that can be used to help create them.

Sometimes people can be dismissive of the inclusion of insights from addictions treatment or the addressing of the distress that immersion in peak oil and climate change awareness can bring about, seeing them as 'touchy-feely' stuff, or as somehow not practical or relevant to actual activism. I would challenge this. I think it is naïve to expect that we could give someone a DVD of *The End of Suburbia* or *An Inconvenient Truth*, which they would then go home and watch on their own, in a darkened house, and expect them to be unaltered by the experience. If finding out about peak oil and really taking on the implications of climate change were distressing for me, it would be reasonable to assume that they might be similarly so for other people too. We see it as entirely reasonable that someone who has gone through a distressing event in

their life might receive counselling afterwards, but is it not logical to think that it might be advisable, given the scale of change our communities are about to undergo collectively, that we might begin to explore now how to, in effect, counsel them through such a transition, and all that it will entail: in effect pre-trauma counselling, rather than post-trauma. I feel that adding this dimension to the work of Transition Initiatives can only strengthen them, and the insights from the application of tools like the FRAMES model open up the possibility of a whole new way of thinking about engaging communities in this work.

Finally, this chapter also explored the first attempt at creating a community-scale vision by students at Kinsale Further Education College in early 2005. Although at the time no-one involved really appreciated the significance of what had been created, it has since become the foundation and the inspiration for the rapidly growing Transition movement. Sometimes an idea appears which is the right idea at the right time, and the Kinsale EDAP is one such idea. Since its publication the idea has evolved and expanded, and become increasingly dynamic. At its heart though is the core message that also underpins the Kinsale plan, that alongside the desire for change, we need to create a vision of where we want to go. The Kinsale EDAP opened with the following quote by Joel Barker: "Vision without action is merely a dream; action without vision just passes the time; vision with action can change the world."[13]

Part Three

THE HANDS

Moving from ideas to action: exploring the Transition model for inspiring local resilience-building

"The time is right to look at what it would mean for the UK over the period of fifteen to twenty years to create a post-oil economy – a declaration less of 'oil independence' and more the end of oil dependence."
– David Miliband, UK Foreign Secretary [1]

"Innovation requires a good idea, initiative, and a few friends." – Herb Shepard [2]

"Let the ideas arise from the community and remain under community control. The job of the Council is to facilitate, to listen, possibly to provide advice, contacts or funds and, most important, to ensure that bureaucracy does not get in the way of grassroots initiatives."
– a local councillor in contact with the Transition Network

Clearly, given the scale of the coming changes outlined above, the idea that we can navigate a safe way through merely by changing our light bulbs and turning the heating down a bit is completely insufficient. In Part Three, 'The Hands', we will look at how we can begin, with the community around us, to move towards a post-oil world that is actually preferable to the present. We stand, potentially, on the cusp of many things, one of which is an unprecedented economic, cultural and social renaissance. The model I have been involved in developing is the Transition model, which is a positive, solutions-focused way of gathering those around you together to start exploring community-scale responses to peak oil and climate change. When we launched the UK's first Transition initiative, Transition Town Totnes, in September 2006, we talked flippantly about the idea 'going viral'. Now, eighteen months later, it has.

The Transition movement has rapidly become one of the fastest-growing community-scale initiatives in the world. In this part of the book, I will attempt to define what a Transition Initiative is, as well as present the Twelve Steps of Transition, the essential ingredients of the early part of a Transition process which should provide what you need to get started. The idea is that by the end of Part Three you will be fully equipped to start this process where you live. The essential message of this part of the book is that we cannot do this as individuals, and that both climate change and peak oil have to underpin both our thinking and our decision making. We need to think bigger, we need to work together with other people and we need very much to accelerate our efforts.

Facing page: planting a cobnut tree in Totnes on a community tree-planting day.

Photo © Teresa Anderson

The Transition concept

So what actually is a 'Transition Initiative'? The initial term used to describe this concept was 'Transition Towns', but this has since become largely irrelevant, given that we are now talking about Transition cities, boroughs, valleys, peninsulas, postcodes, villages, hamlets and islands . . . So although none of these alliterates quite as nicely as Transition Towns, Transition Initiatives seems to be the best overall term. Transition Initiatives are an emerging and evolving approach to community-level sustainability, which is starting to appear in communities up and down the country. They are, to use a term coined by Jeremy Leggett, "scalable microcosms of hope". The idea began, as we have seen in Chapter 9, with the Kinsale Energy Descent Action Plan in Ireland, and has since spread to communities around the UK and beyond.

Transition Initiatives are based on four key assumptions:

1) That life with dramatically lower energy consumption is inevitable, and that it's better to plan for it than to be taken by surprise.

2) That our settlements and communities presently lack the resilience to enable them to weather the severe energy shocks that will accompany peak oil.

3) That we have to act collectively, and we have to act now.

4) That by unleashing the collective genius of those around us to creatively and proactively design our energy descent, we can build ways of living that are more connected, more enriching and that recognise the biological limits of our planet.

The future with less oil could, if enough thinking and design is applied sufficiently in advance, be preferable to the present. There is

Conventional Environmentalism	The Transition Approach
Individual behaviour	Group behaviour
Single issue	Holistic
Tools: lobbying, campaigning and protesting	Tools: public participation, eco-psychology, arts, culture and creative education
Sustainable development	Resilience/relocalisation
Fear, guilt and shock as drivers for action	Hope, optimism and proactivity as drivers for action
Changing National and International policy by lobbying	Changing National and International policy by making them electable
The man in the street as the problem	The man in the street as the solution
Blanket campaigning	Targeted interventions
Single level engagement	Engagement on a variety of levels
Prescriptive – advocates answers and responses	Acts as a catalyst – no fixed answers
Carbon footprinting	Carbon footprinting plus resilience indicators
Belief that economic growth is still possible, albeit greener growth	Designing for economic renaissance, albeit a local one

Figure 18: How the Transition approach is distinct from other environmental approaches.

no reason why a lower-energy, more resilient future needs to have a lower quality of life than the present. Indeed, a future with a revitalised local economy would have many advantages over the present, including a happier and less stressed population, an improved environment and increased stability.

In fact some awareness of this is starting to appear at government level, in Australia at least. Andrew McNamara, Queensland's Minister for Sustainability, Climate Change and Innovation recently said:

> "There's no question whatsoever that community-driven local solutions will be essential. That's where government will certainly have a role to play in assisting and encouraging local networks, who can assist with local supplies of food and fuel and water and jobs and the things we need from shops. . . .
>
> We will see a relocalisation of the way in which we live that will remind us not of last century, but the one before that. And that's not a bad thing. Undoubtedly one of the cheaper responses that will be very effective is promoting local consumption, local production, local distribution. And there are positive spinoffs to that in terms of getting to know our communities better. There are human and community benefits from local networks that I look forward to seeing grow."[3]

How this is explored and developed in practice will be different in each settlement: rather than offering prescriptive solutions, Transition Initiatives aim to act as catalysts for a community to explore and come up with its own answers.

They focus the collective mind on the practicalities of energy descent, which, as we saw in Part One, an increasing number of commentators argue will be the inevitable outcome of both peak oil and climate change. Transition

TRANSITION INITIATIVES IN ORDER OF ADOPTION (continued)

All places are in the UK except for those shown in italics.

Waiheke Island, New Zealand
Orewa, New Zealand
Dunbar, Scotland
Rhayader, Wales
Seaton, England
Bath, England
Exeter, England
Isle of Man, UK
Canterbury, England
Kapiti District, New Zealand
Carbon Neutral Biggar, Scotland
Presteigne, Wales
Wolverton, England
Leicester, England
Holywood, Northern Ireland
Westcliff-on-Sea, England
Isles of Scilly, UK
Liverpool South, England
Norwich, England
Tring, England
Crediton, England
Boulder, CO, USA
North Howe, Scotland
Lampeter, Wales
South Petherton, England
Armidale, NSW, Australia
Chichester, England
Bell, VIC, Australia
Bellingen, NSW, Australia
Berkhamsted, England
Forres, Scotland
Sandpoint, ID, USA
Opotiki Coast, New Zealand
Newcastle, NSW, Australia
Chepstow, Wales
Coventry, England
Bungay, England
Nelson, New Zealand
Hervey Bay, QLD, Australia
Dorchester, England
New Forest, England

Initiatives aim to create communities that are resilient; that is, more able to withstand shocks from outside (a concept explored in Chapter 3), be they from climate change, problems of energy security, or rising fuel prices. Rather than being just an intellectual exercise, they explore the practicalities of the conscious relocalisation of a settlement; to paraphrase David Holmgren (talking about permaculture, but the same is true in this context), Transition Initiatives are 'the wholehearted and positive acceptance of energy descent, as not only inevitable but as a desired reality'.[4]

Given that oil and gas are depleting resources, and that we urgently need extreme cuts in CO_2 emissions, even to the extent that our daily lives sequester more carbon than they produce, Transition Initiatives ask what would such a world actually look like? How would we live? Where would our food come from? What would we hear when we opened the window in the morning? The Transition process offers a positive, solutions-focused approach that draws together the various elements of a community to address this common challenge and sees much of the solution as coming from within, through a process of unlocking what is already there, rather than from experts and consultants coming in from the outside.

In Figure 18 on page 135 I attempt to distinguish how the Transition approach differs from more conventional approaches to environmentalism, having put resilience-building as one of its key objectives. I appreciate that in my 'Conventional Environmentalism' column I have, to an extent, set up a straw horse, so generalised that it verges on a stereotype, but I think this process is an important one, essential in distinguishing the distinctive ground that the Transition approach stands on.

The philosophical underpinnings

One of the principal foundations of the Transition concept is permaculture. Permaculture is something notoriously difficult to explain in a single sentence: it resists an off-the-cuff definition that would enable an accurate mental picture to be formed. In essence, it is a design system for the creation of sustainable human settlements. When designing the transition that our settlements and communi-

Integrating chickens with vegetable production has long been a holy grail of permaculturists. Here at Dominic Waldron's permaculture smallholding in Co. Kerry, Ireland, chickens conduct slug patrol in between the vegetables which are protected from their affections by hoops and fruit cage netting.

The Ark Permaculture project near Clones, Co. Monaghan, Ireland – a wonderful example of permaculture principles in practice. Note how intelligent siting of the pond allows it to reflect winter light into the house, warm the vegetable garden, provide water and nutrients for irrigation, as well as passive cooling for the house. Siting the vegetable garden just outside the back door makes for easier maintenance and optimises the microclimate created by the house. Without permaculture design these three elements – pond, house and garden – would have been assembled randomly and these beneficial relationships lost.

ties will inevitably have to undertake, we need a design template with which we can successfully assemble its various components – social, economic, cultural and technical – in the most efficient way possible. Permaculture can be thought of as the design 'glue' and the ethical foundations we use to underpin Transition work, to stick together all the elements of a post-peak settlement. The reason that people with a permaculture background tend to 'get' the Transition concept ahead of most other people is that it is based on permaculture design principles.[5] I have spent the last ten years teaching permaculture, and its ethics and principles very much underpin my thinking.

Permaculture was originally conceived in the 1970s at the time of the first oil crises, as being a 'Permanent Agriculture', moving away from annual cropping and monoculture in agriculture to multi-layered systems making use of productive and useful trees and perennial plants.[6] Its focus on agricultural systems soon broadened, as it became clear that sustainability in food cannot happen in isolation from the

David Holmgren's Permaculture Principles

(adapted from Permaculture: principles and pathways beyond sustainability)

"Permaculture is a design system based on ecological principles which provides the organising framework for implementing a permanent or sustainable culture. It . . . draws together the diverse skills and ways of living which need to be rediscovered and developed to empower us to move from being dependent consumers to becoming responsible producers.

In this sense, permaculture is not the landscape, or even the skills of organic gardening, sustainable farming, energy-efficient building or eco-village development as such, but can be used to design, establish, manage and improve these and all other efforts made by individuals, households and communities towards a sustainable future."

– David Holmgren (2002),
Permaculture: Principles and Pathways Beyond Sustainability.
Holmgren Design Services

"Permaculture is a philosophy of working with, rather than against nature; of protracted and thoughtful observation rather than protracted and thoughtless labour; and of looking at plants and animals in all their functions, rather than treating any area as a single-product system."

– Bill Mollison

1. Observe and Interact

The power of good observation is something not many of us have, and detailed observation of where we are will underpin any actions we undertake. A post-peak world will depend on detailed observation and good design rather than energy-intensive solutions.

2. Catch and Store Energy

Energy passes through our natural systems, and is stored in a variety of ways, in water, trees, plants, soils, seeds and so on. We need to become skilled at making best use of these, and move our idea of 'capital' from what we have in the bank, to the resources we have around us. I once heard Holmgren say that a good woodpile, such as you would see in Eastern Europe, is a far more reasonable indicator of national wealth than GDP.

3. Obtain a Yield

This principle states that any intervention we make in a system, any changes we make or elements we introduce ought to be productive, e.g. productive trees in public places, edible roof gardens, or urban edible landscaping.

4. Apply Self-regulation and Accept Feedback

A well-designed system using permaculture principles should be able to self-regulate, and require the minimum of intervention and maintenance, like a woodland ecosystem, which requires no weeding, fertiliser or pest control.

5. Use and Value Renewable Resources and Services

Where nature can perform particular functions, be it aerating soil (worms), fixing nitrogen (clover) or building soil (trees) we should utilise these attributes, rather than thinking we can replace them. Where nature can take some work off our hands we should let it.

6. Produce No Waste

The concept of waste is essentially a reflection of poor design. Every output from one system could become the input to another system. We need to think cyclically rather than in linear systems.

7. Design from Patterns to Details

We need to be able to keep looking at our work from a range of perspectives. This principle argues that we need to see our work in the wider context of watershed, regional economy and so on, so as to keep a clearer sense of the wider canvas on which we are painting, and the forces that affect what we are doing.

8. Integrate Rather Than Segregate

Permaculture has been described as the science of maximising beneficial relationships. In a powered-down settlement, what will become increasingly important is the relationships that we can weave between different elements of the place (a principle demonstrated in Tool 5, p.101). Solutions are to be found in integrated wholistic solutions rather than increased specialisation and compartmentalisation.

9. Use Small and Slow Solutions

This principle represents the core argument of this book, that, as Holmgren puts it, "systems should be designed to perform functions at the smallest scale that is practical and energy-efficient for that function." Our solutions will be based on the principle that the smaller and more intensive they can be, the more resilient they will be.

10. Use and Value Diversity

Monocultures are incredibly fragile and prone to disease and pests, more diverse systems have much more inbuilt resilience. Our towns will be much more able to prosper during energy descent if they have a diversity of small businesses, local currencies, food sources, energy sources and so on than if they are just dependent on centralised systems, globalisation's version of monoculture.

11. Use Edges and Value the Marginal

One of the observations used a lot in permaculture is the idea of 'edge', that the point where two ecosystems meet is often more productive than either of those systems on their own. This principle reminds us of the need to overlap systems where possible so as to maximise their potential.

12. Creatively Use and Respond to Change

Natural systems are constantly in flux, evolving and growing. The way they respond to shock, such as forest fires, can teach us a great deal about how we might manage the transition away from fossil fuels. Remaining observant of the changes around you, and not fixing onto the idea that anything around you is fixed or permanent will help too.

> "Permaculture is the conscious design and maintenance of agriculturally productive systems which have the diversity, stability, and resilience of natural ecosystems. It is the harmonious integration of the landscape with people providing their food, energy, shelter and other material and non-material needs in a sustainable way."
>
> – Graham Bell

range of other elements that make up society – economics, building, energy and so on. The term 'permaculture' became seen as a contraction of 'permanent culture', being about the creation of a culture of permanence. Its most thorough early exposition, Bill Mollison's *Permaculture: a Designer's Manual,*[7] was, in effect, a manual for Earth repair, an astonishingly broad, ambitious, encyclopaedic work which offered the reader a toolkit for Earth restoration. Over the next fifteen years permaculture, at least in the mainstream psyche (despite growing massively and inspiring and underpinning thousands of projects around the world), became perceived by many as an odd form of gardening using car tyres and obscure plants which probably no one would want to sit down to for supper.

In 2004, David Holmgren, the co-originator of the concept, published *Permaculture: Principles and Pathways Beyond Sustainability,* which put permaculture back on the map as a radical design science, and redefined the principles of permaculture as the principles that will be needed to underpin a post-peak world.

When I encountered peak oil, my first response was instinctively to utilise permaculture principles when formulating a response. It struck me that the movement I was so fond of and a part of was still at such a small point in its development in terms of its prominence in national awareness, and the need for its insights in informing a massive-scale social transformation, that we needed to really ramp things up hugely. I began to wonder why that was. Then I came across an excellent and insightful article by Eric Stewart, in which he wrote:

> "It seems to me that permaculture houses two virtually polar impulses: one involves removal from larger society; the other involves working for the transformation of society. While the case

"All things are possible once enough human beings realise that everything is at stake."

– Norman Cousins

A new housing development under construction near Mallow, Co. Cork, Ireland. This typical development shows what happens when permaculture principles are absent. If the main building here is south-facing (which would reduce its heating requirements by up to 15%), then the house to its right receives sunshine mainly on its gable end. Also the land around the houses has not been intentionally designed – it is just land left over after building. In this part of Ireland, cob building was a traditional form of construction, with perfect soil for cob, yet here the houses are all made from distantly produced concrete while the cob soil awaits disposal.

> can be made that removal from the larger society represents action that is transformative of society, I believe that there is an imbalance within the cultural manifestation of permaculture that has favoured isolation over interaction. The cultural shift we need depends on increasing interaction to increase the availability of the resources permaculture offers."[8]

This hit the nail on the head for me. Permaculture is a movement which offers, as redefined by Holmgren, the design system and philosophical underpinning of a post-peak society, yet at the same time, according to Stewart, it is often guilty of maintaining a distance from that society. Peak oil, to me, is a call to the bodgers and chairmakers in the woods, the market gardeners and orchardists up misty rural lanes, the small-scale wind installers on the windswept highlands, to bring all the wonderful skills they have accumulated, the insights they have obtained through years of practice and contemplation, back to where the mass of the population is starting to realise things are not right. It is a call to learn new ways of communicating with the mainstream, and with an ethic of service, to seek to engage with others on an unprecedented scale.

The Transition approach is, I hope, one in which permaculture principles are implicit, not explicit. It is my attempt to get round the fact that permaculture is a concept that is very hard to explain to the person in the pub who asks you what it means, if you don't have a flip-chart and pens and fifteen minutes in which to draw pictures of chickens and ponds and greenhouses. Permaculture principles underpin this approach, which is designed to mainstream its concepts, presenting them as fundamental to any response to energy descent. Yet somehow the concept of Transition is easier to explain, allowing more time for other conversations. So, if you have a background in permaculture, and some of this feels familiar, that's why.

Six principles that underpin the Transition model

There are six principles that I feel define what is distinctive about the Transition concept. They have emerged from observing the process as it has unfolded, and, I think, neatly sum up what is unique to this evolving approach.

1. Visioning

The concept of visioning was explored in depth in Chapter 7. In the context of these six principles, visioning refers to the fact that the Transition approach has, as a fundamental principle, the belief that we can only move towards something if we can imagine what it will be like when we get there. The vision we have in our mind when we set out on this work will go a long way towards determining where we will end up. Are we working towards Holmgren's 'Techno Explosion' (see p.46), or perhaps something more realistic and desirable? Creating a clear and enticing vision of our desired outcome is a key principle of the Transition process.

2. Inclusion

The scale of the challenge of peak oil and climate change cannot be addressed if we choose to stay within our comfort zones, if 'green' people only talk to other 'green' people, business people only talk to other business people, and so on. The Transition approach seeks to facilitate a degree of dialogue and inclusion that has rarely been achieved before, and has begun to develop some innovative ways of bringing this about. This is seen as one of the key principles simply because without it we have no chance of success.

3. Awareness-raising

The end of the Oil Age is a confusing time. We are constantly exposed to bewildering mixed messages. The media presents us with headlines such as "Steep decline in oil production brings risk of war and unrest, says new study",[9] and "Carbon output rising faster than forecast, says study",[10] yet at the same time advertising puts across the conflicting message that business as usual is the only way forward, that globalisation is the only model that can feed the world, and that just buying this next thing will make us happy. Indeed the contrast can sometimes be striking, with an article about the melting of Arctic ice-sheets next to an advert for a new car or cheap flights.

The media to which we are increasingly exposed continually give out double messages,

"Our first task is to create a shadow economic, social and even technological structure that will be ready to take over as the existing system fails."

– David Ehrenfeld

which can leave one feeling perplexed. Sometimes new Transition Initiatives feel that they don't need to do much awareness-raising because everyone must be aware of these issues by now, but it is essential to start with the assumption that people don't know anything about these issues. We need to assume no prior knowledge, and set out the case as clearly, accessibly and entertainingly as possible, giving people the key arguments in order to let them formulate their own responses.

4. Resilience

In Chapter 3 we explored the concept of resilience, but it is useful to restate at this point that the rebuilding of resilience is, alongside the need to move rapidly to a zero carbon society, central to the Transition concept. Indeed, to do one without the other will fail to address either challenge.

5. Psychological insights

Insights from psychology are also key to the Transition model. It is understood that among the key barriers to engagement are the sense of powerlessness, isolation and overwhelm that environmental issues can often generate.[11] These do not leave people in a place from which they can generate action, either as an individual or as a community. The Transition model uses these insights firstly through the creation of a positive vision (see Principle 1, p.141), secondly by creating safe spaces where people can talk, digest and feel how these issues affect them, and thirdly by affirming the steps and actions that people have taken, and by designing into the process as many opportunities to celebrate successes as possible. This coming together – the sense of not being the only person out there who is aware of peak oil and climate change and who finds it scary – is very powerful. It enables people to feel part of a collective response, that they are part of something larger than themselves.

6. Credible and appropriate solutions

In the film *Way to Go: Life at the End of Empire*,[12] Tim Bennett talks about what he calls the 'happy chapter' at the end of most environmental books, which spend nine chapters telling you how dreadful everything is, and one on the end with a few token solutions. Similarly, I have heard many a talk where the speaker has set out the scale of the climate challenge, and at the end has one slide about turning down our thermostats and changing our light bulbs.

It is important that Transition Initiatives, having laid out the peak oil and climate change arguments, enable people to explore solutions of a credible scale. One of the reasons behind what we might call the 'light-bulb syndrome' is that people are often only able to conceive two scales of response; individuals doing things in their own homes, or the government acting on a national scale. The Transition model explores the ground between these two: what could be achieved at a community level.

The Project Support Project concept

One of the things that distinguishes the Transition approach is the concept of the Project Support Project (PSP). Ideally, we need Transition Initiatives to be self-organising, and able to harness the passion and enthusiasm that the process unleashes. Whilst looking around for such models, I happened by some considerable serendipity to meet with John Croft of the Gaia Foundation of Western Australia.[13] A couple of months later he returned and ran a one-day training course on his approach for Transition Town Totnes. Some of the tools he had developed, in particular the approach he calls 'Dragon Dreaming', can be found in Appendix 2, but the concept of most relevance here is that of the Project Support Project.

The Gaia Foundation has catalysed and supported hundreds of projects, and has done much work in developing organisational models. It is a small group that has no one person at the centre, and that is founded on a set of shared principles. Any project supported by the Foundation agrees with the following:

1. It involves the personal growth of those involved

2. It strengthens and/or builds community

3. It works in service of the Earth.

Any projects that meet these criteria (Croft recommends no more than six) can apply to become a Gaia Foundation project. Each project has its own bank account, makes its own decisions, and so on. In essence, the concept of a PSP is that, rather than being an organisation that co-ordinates and drives a wide range of projects itself, the aim is instead to create an atmosphere within which projects emerge and then to support them when they do. This means that the organisation can be much lighter and more responsive and, in effect, truly act as the catalyst that these projects are intended to be.

With Transition Town Totnes, we have made this a central concept. We see the role of TTT as an organisation to raise awareness, to continuously raise the profile of the project and its aims, to build interest in the concepts as a whole, and to build enthusiasm for the Transition 'brand'. We exist to inspire and motivate the initiation of projects, and then to network and nurture them once they start. Within this model, one has to be careful that the integrity of the name is preserved. In order for

someone to call a project they are doing a TTT project, they need to submit an A4 sheet outlining their proposal. One example is the Book and DVD Resource project, where a woman in Totnes decided she would like to create a collection of sustainability-related books and films, the books being available in the library and the DVDs for free hire in the DVD shop. She had decided that she wanted to do this, she asked for endorsement as a TTT project, which she got, and there are now £1,500 worth of books in the library that otherwise wouldn't be there.

One of Croft's suggestions is that groups ask themselves an important key commitment question. Once their proposed project has been effectively worked into a plan, and the draft budget finalised the planning group considers this: "If this project would not get funding from elsewhere, would those involved be willing to carry the financial burden of any losses incurred by the project?" The group driving the second launch of the Totnes Pound found this very useful, a real focuser of minds and generator of commitment.

Issues of scale

One of the questions we are often asked is what is the ideal scale for a Transition Initiative. In many ways market towns, which are on the scale that many of the first Transition Initiatives started, are the ideal scale. They have a clear hinterland, historically defined by the villages and rural areas whose inhabitants brought their produce to that town rather than to an adjoining one. Similarly, islands are a good scale to work on, as they have a clearly defined boundary. Why the concept of 'Transition Towns' felt so right at the beginning was that the small town is a scale we can all innately relate to. Many people living in a large city crave the

"What it takes is a scale at which one can feel a degree of control over the processes of life, at which individuals become neighbours and lovers instead of just acquaintances and ciphers, makers and creators instead of just users and consumers, participants and protagonists instead of just voters and taxpayers. That scale is the human scale."

– Kirkpatrick Sale (1980),
Human Scale,
Coward, McCann and Geoghegan

more identifiable scale of a town, or in this context, the neighbourhood. Many people feel that as globalisation has increased, the sphere that we are connected to and can actually influence has shrunk. Perhaps so few people vote now because they have come to feel that their vote makes no difference.

I have come to think that the ideal scale for a Transition Initiative is one over which you feel you can have an influence. A town of 5,000 people, for example, is one that you can relate to; it is one with which you can become familiar. Having grown up in Bristol, I am aware that most cities were, historically, a collection of villages, and still have that feel to them. This concept of working at a neighbourhood scale is not a new one.[14]

Ultimately, you will get a sense of what is the optimal scale for your initiative. Indeed, you will probably instinctively already have a sense of this. As you look around you, what feels like the optimal scale to be working on? Where, instinctively, do you feel your sphere of influence to be? Transition Bristol, the first city-scale initiative, seeks to network, inspire, train and enable, and to support the emerging neighbourhood-scale initiatives, Transition Redland, Transition Withywood and so on, in their own Transition Initiative.[15]

There is no magic formula for the question of scale. Your group will need to follow its instincts, but don't worry about it – it will emerge naturally. Do resist the temptation, which has arisen for some, to try to start too big, thinking at the scale of Transition Yorkshire, or Transition Scotland. While useful as concepts, they are really putting the cart before the horse. While it may be the case that at some point in the future the broad spectrum of groups in a geographical area may recognise a need to network themselves to maximise their effectiveness, this needs to grow from a base of a network of vibrant Transition communities, rather than be created in advance (you can see how the Transition Network encourages groups at different scales in Appendix 5, 'How to become a Transition Initiative').

The interface between Transition Initiatives and local politics

The power of the Transition process is its potential to create a truly community-led process which then interfaces with local politics, but on its own terms. The role we identify for Local Authorities in this process is to support, not to drive it. Local Agenda 21, although it created many interesting initiatives, was in essence a top-down process trying to pretend that it wasn't. It is important that Transition Initiatives operate independently of input from local politicians, at least to begin with. A Transition Initiative could not, by definition, be a project conceived and driven forward by a Council, although it is one where the active and enthusiastic support of local government is invaluable. What has been happening increasingly in recent months is that the first contact from a community is from someone in the local council, be it District, Parish or Town Council. Sometimes a Council member will end up as part of the Steering Group, or the Council will offer their support in a range of ways.

In the book *Peak Oil Prep*, Mick Winter argues that one of the main roles of state government in the US (for which read national government for the UK) is to "stay out of the way of local governments". He writes:

"They know better than the state what they need. Give them whatever they want. Focus on projects that serve regions . . . If something can be done at the local level, states should give communities what they need to make it happen – with no strings. If there's something that can only be done at the state level, then that is the state's responsibility." [16]

In the UK one can extend this model down another level, and say that the role of local government is also to facilitate Transition processes, not to lead or guide them but to support them. Increasingly the Councillors who get in touch do have an understanding of this process, and are actively seeking to help facilitate it. One Chairman of a Town Council who contacted the Transition Network wrote:

"Whilst I would see the Council being supportive to a Transition movement, one of the things that I found most attractive about Transition Initiatives was the grass-roots community involvement. In my experience the very best model is the Council supporting and encouraging the various communities, but much if not most of the initiative coming from the various community groups.

We as councillors need to be aware that Transition Initiatives are not something that we bestow on the community; it is not going to be just a badge or symbol for the Council, it is something that will happen anyway. Though Council support will help and assist the birth, the Council may also help the ideas to move into parts of the community that might otherwise not be reached."

When Transition Initiatives do approach their local or district council, they do so representing a significant part of the community, and with a groundswell of momentum behind them. In Kinsale, once the EDAP was done, a motion endorsing it was submitted to Kinsale Town Council and unanimously approved. In Totnes, six months after the Official Unleashing, the Council passed a resolution endorsing the work of Transition Town Totnes (TTT). This support is very powerful in terms of being able to drive the initiative forward with enhanced credibility, but should only be sought once the project has an established track record and has forged its own identity.

For many towns in the US, such as Portland and Oakland, [17] the passing by the local authority of a 'Peak Oil Resolution' is seen as a key step. This may be the case, but my sense is that the important first steps are to engage the community in the awareness-raising and building the energy for the project, rather than disappearing at an early stage into the bewildering world of policy writing and working at the local government level. Once you have achieved this, local government will want to be part of the process because they can see it as being where the energy and innovative thinking is taking place.

In terms of TTT's interaction with the local authority, one of the most important elements of this is its Liaison with Local Government Group. This was formed by a group of people who had been involved for some time as local councillors, or had sat on various bodies and understood how the political structure works.

This group goes through each new programme of events that is coming up and invites the public representatives who they feel should be there. They also keep an eye on upcoming council consultations. They are a centrally important part of the TTT Initiative. One could argue that if at an early stage prominent local political figures want to get involved, their role is to work with such a group to drive forward the whole larger process.

Chapter II

How to start a Transition Initiative

At this stage you might, hopefully, be thinking that you would like to start a Transition Initiative in your community. You might be looking around you and wondering where to start, how on earth you might be able to even begin planning such an Initiative. Below I will introduce the Twelve Steps of Transition, which will hopefully address your 'Where do we start, and then what?' questions.

Before that, though, it is useful to address some of the questions that often arise for people at the early stage of planning a Transition Initiative, and which may well prevent them from proceeding any further. I call these 'The Seven Buts'.

The Seven 'Buts'

"BUT . . . We've got no funding"

This really is not an issue. Funding is a very poor substitute for enthusiasm and community involvement, both of which will take you through the first phases of your transition. Funders can also demand a measure of control, and may steer the Initiative in directions that run counter to community interests and to your original vision. It should be straightforward for your Initiative to generate an adequate amount of income. Transition Town Totnes began in September 2005 with no money at all, and has mostly been self-funding until recently. The talks and film screenings that we run bring in money to subsidise free events such as Open Space Days. You will reach a point where you have specific projects that will require funding, but until that point you'll manage. Retain the power over whether your important Initiative happens, and don't let lack of funding stop you.

"BUT . . . They won't let us"

There is a fear among some green folks that somehow any Initiative that actually succeeds in effecting any change will get shut down, suppressed, attacked by faceless bureaucrats or corporations. Transition Initiatives operate 'below the radar'; as such, they don't incur the wrath of any existing institutions.

On the contrary, with corporate awareness of rising energy prices and climate change building daily, you will be surprised at how many people in positions of power will be enthused and inspired by what you are doing, and will support, rather than hinder, your efforts. You will find your Transition Initiative is constantly pushing on open doors. The unanimous endorsements of many Transition Initiatives by their local councils is one example of this.

"BUT . . . There are already green groups in this town, and I don't want to step on their toes"

We'll go into this in more detail in Step 3 (page 152), but in essence, you'd be exceedingly unlucky to encounter any 'eco-turf wars'. What your Transition Initiative will do is form a common goal and sense of purpose for the existing groups, some of which you might find are a bit burnt out and will really appreciate the new vigour you will bring. Liaising with a network of existing groups towards an Energy Descent Action Plan will enhance and focus their work, rather than replicate or supersede it. Expect them to become your allies, crucial to the success of your Transition process.

"BUT . . . No one in this town cares about the environment anyway"

One could easily be forgiven for thinking this, given the existence of what we might perceive as an apathetic consumer culture surrounding us. Scratch a bit deeper though, and you'll find that people are already passionate about many aspects of what Transition Initiatives will focus on. The most surprising of people are keen advocates of key elements of a Transition Initiative – local food, local crafts, local history and culture. The key is to go to them, rather than expecting them to come to you. Seek out common ground, and you'll find your community to be a far more interesting place than you thought it was.

"BUT . . . Surely it's too late to do anything?"

It may be too late, but the likelihood is that it isn't. Your (and others') endeavours are absolutely crucial. Don't let hopelessness sabotage your efforts. It is within your power to maximise the possibility that we can get through this – don't give that power away.

"BUT . . . I don't have the right qualifications"

If you don't do this, who else will? It matters not that you don't have a PhD in sustainability, or years of experience in gardening or planning. What's important is that you care about where you live, that you see the need to act, and that you are open to new ways of engaging people.

Useful qualities for someone starting a

Transition Initiative are:

- Positive
- Good with people
- A basic knowledge of the place and some of the key people in the town.

That, in truth, is about it. You are, after all, about to design your own demise into the process from the start (see Step 1 overleaf), so your role at this stage is like a gardener preparing the soil for the ensuing garden, which you may or may not be around to see.

"BUT . . . I don't have the energy for doing that!"

As the quote often ascribed to Goethe goes, "Whatever you can do or dream you can, begin it. Boldness has genius, power and magic in it!" The experience of beginning a Transition Initiative certainly shows this to be the case. While the idea of preparing your town (or city, hamlet, valley or island) for life beyond oil may seem staggering in its implications, there is something about the energy unleashed by the Transition process that is unstoppable.

Everyone I have spoken to who has initiated a Transition project, has had a period after a few weeks of thinking, "What have we started here?!" It may feel that you will have to do it all yourself. You may feel overwhelmed by the prospect of all the work and complexity, but people will come forward to help. Indeed, many have commented on the serendipity of the whole process, how the right people appear at the right time. There is something about seizing that boldness, about making the leap from 'why is no-one doing anything' to 'let's do something', that generates the energy to keep it moving.

Very often, developing environmental initiatives feels like pushing a broken-down car up a hill; a hard, unrewarding slog. Working with a Transition Initiative often feels like coming down the other side – the car starts moving faster than you can keep up with, accelerating all the time. Once you give it that push from the top of the hill it will develop its own momentum. That's not to say it isn't hard work sometimes, but it is almost always a pleasure.

The Twelve Steps of Transition

These Twelve Steps emerged from observing how the Transition Town Totnes initiative evolved, and from other communities contacting us to ask what we were doing. They don't take you from A-Z, rather from A-C, which is as far as we've got with this model so far. These steps don't necessarily follow each other logically in the order they are set out here; every Transition initiative weaves a different way through the Steps, as you will see. These Twelve Steps are still evolving, in part shaped by your experience of using them. There may end up being as few as six or more than fifty!

It is important to observe that they are not meant to be prescriptive; rather they are intended to suggest pieces of the puzzle you may choose to assemble. You do not have to follow them religiously, step-by-step: you can use the ones that feel useful, discard the ones that don't, and add in new ones that you come up with. As you will see, many of the communities that have already started this process have already begun assembling them in different ways.

1. Set up a steering group and design its demise from the outset

Bill Mollison, the co-originator of the permaculture concept, once famously said, "I can't save the world on my own. It'll take at least three of us", or words to that effect. In starting your Transition Initiative you will need to gather some like-minded souls in order to drive forward the first stage of the process. What is essential though, and its importance is becoming increasingly clear, is that from its first meeting, that group must design its own demise, set a defined lifespan for its functioning.

So many groups get atrophied and stuck with people who cling to their roles in a way that stifles the progress of the project. In the longer term it is important that the project becomes driven by those who are actually doing things. I would suggest that you form your Steering Group of reliable people with the aim of getting through Steps 2 to 5, and agree that once a minimum of four sub-groups are formed, your group disbands and the Steering Group becomes made up of one person from each of the groups. This requires a degree of humility, but is very important in order to put the success of the project above the individuals involved. It is also quite a relief! It means that you aren't forming a group whose aim is the complete relocalisation of the settlement in question; just to do the first few Steps – a much more manageable task!

Transition Tip

- *Invite a professional facilitator/change manager to help you*
- *Involve everybody in the transformation*
- *Create, together, clear written aims and principles for the new formation (see our Wiki) and refer to them frequently*
- *Try to stay unattached to outcome, and let go of your own agendas*
- *Some people will leave and others will join – whoever turn up are the right people*
- *Trust the process! A new mindset takes time to take root*

(by Adrienne Campbell, Transition Town Lewes)

2. Raise awareness

You cannot assume that people in your community are familiar with peak oil, climate change, or even with basic environmental concepts and principles that you might take for granted. It is essential before launching an Official Unleashing event (see 4 below) that you have prepared the ground. In Totnes we spent nearly a year giving talks, film screenings and networking before we organised the launch. During that time we learned a great deal about how to do this most effectively.

We screened *The End of Suburbia* three times, and had a full room and a completely different audience each time. Various methods for facilitating film screenings can be read about in Tool for Transition no. 7 (see p.154). Other films we showed were *The Power of Community* and *Peak Oil: Imposed by Nature* (other ideas for films can be found in the Resources section). One important point is that you can never assume that everyone has seen the films and that no one will come if you show them again. These films create a ripple effect and lots of people want to see them. It is important that these screenings are presented in such a way that they are fun and memorable, and create a buzz, so that people go home and tell their friends and family.

Transition Tip

- *Start any film screening or talk by inviting people to turn to the person next to them and tell them who they are, where they have come from, and why they are here. Then after the film (or talk) do the same thing (but with the person on their other side), this time to talk about their thoughts on the film. People enjoy doing this, it really enhances their enjoyment of the evening and is a powerful tool for starting to build connections.*

The first poster for Transition Town Totnes

Another aspect of this awareness-raising work is talks. It is essential to avoid a series of peak oil talks which are doom-laden evenings about how civilisation is about to implode and we are all about to start eating each other when oil hits $120 a barrel. This will lead to your Transition Initiative falling at the first hurdle. Find speakers who can present the matter in a positive, engaging way. Organise events that make people think, but which also support them through the process of realising the illusory nature of the oil-created world around them, which for some can be quite traumatic. You need to be prepared for the diverse manifestations of 'post-petroleum stress disorder'. Make sure you design enough space into your events for peo-

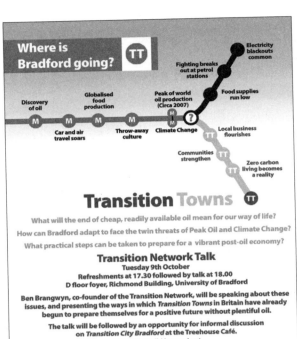

Where is Bradford going? TT

Discovery of oil — Globalised food production — Peak of world oil production (Circa 2007) — Fighting breaks out at petrol stations — Food supplies run low — Electricity blackouts common

Car and air travel soars — Throw-away culture — Climate Change — Local business flourishes — Communities strengthen — Zero carbon living becomes a reality

Transition Towns

What will the end of cheap, readily available oil mean for our way of life?

How can Bradford adapt to face the twin threats of Peak Oil and Climate Change?

What practical steps can be taken to prepare for a vibrant post-oil economy?

Transition Network Talk
Tuesday 9th October
Refreshments at 17.30 followed by talk at 18.00
D floor foyer, Richmond Building, University of Bradford

Ben Brangwyn, co-founder of the Transition Network, will be speaking about these issues, and presenting the ways in which *Transition Towns* in Britain have already begun to prepare themselves for a positive future without plentiful oil.

The talk will be followed by an opportunity for informal discussion on *Transition City Bradford* at the Treehouse Café.
(Food will be available to purchase)

Please call 01274 732354 for more details, or call into Treehouse, Bradford Centre for Nonviolence, 2 Ashgrove, Bradford BD7 1BN.
visit www.treehousecafe.org for a full list of *Transition City Bradford* awareness events

treehouse · beat · centre for Community Engagement · UNIVERSITY OF BRADFORD MAKING KNOWLEDGE WORK

TRANSITION TOWN BRIXTON
www.transitiontownbrixton.org
presents an inspiring film about the benefits of localisation
- followed by Q&A -

THE POWER OF COMMUNITY
How Cuba Survived Peak Oil

When the Soviet Union collapsed in 1990, Cuba's economy went into a tailspin. With imports of oil cut by more than half – and food by 80 percent – people were desperate. This film tells of the hardships and struggles as well as the community and creativity of the Cuban people during this difficult time. Cubans share how they transitioned from a highly mechanized, industrial agricultural system to one using organic methods of farming and local, urban gardens. It is an unusual look into the Cuban culture during this economic crisis, which they call "The Special Period." The film opens with a short history of Peak Oil, a term for the time in our history when world oil production will reach its all-time peak and begin to decline forever. Cuba, the only country that has faced such a crisis – the massive reduction of fossil fuels – is an example of options and hope.
For more information and trailer visit www.powerofcommunity.org

Monday 8 October, 7pm, Southside Bar, Free
(formerly the George IV) 144 Brixton Hill SW2 1SD. Happy hour till 8pm. Food till 10pm.

GROWING
14 April 2007
10–4pm

YOUR OWN

Venue:
Lewes New School
Talbot Terrace
Lewes

Led by Nir Halfon
Biodynamic horticulturalist and permaculture designer

Bring vegetarian lunch to share

Price £20 – ring 0781 5035262 to book a place

A one-day workshop for the complete novice

FOOD

Some of the areas covered are:
Growing fruit and vegetables in every situation
Composting
Pests and weeds

EARTH · Earth Art skills – skills for sustainable living

In association with TRANSITION TOWN LEWES
www.transitiontowns.org/Lewes
Unleashing 24 April 2007

TF Resources Database:
FLI (Falmouth Living Information)

Transition Falmouth

Please tell us what is going on in Penryn, Falmouth and the surrounding area for:

Local Youth, Food and Health

email details to:
fli@transitionfalmouth.org.uk
or write to:
FLI c/o Transition Falmouth
45 Grenville Road, Falmouth TR11 2NP

For more information :
www.transitionfalmouth.org.uk
Lorely 01326 317587

A selection of posters from Transition Initiatives

A Seedy Saturday
The first Falmouth Community Seed Swap

9:30am-12:30pm 21st April
Falmouth Women's Institute, Webber Street

An exciting opportunity to meet people interested in gardening, local food production and sustainability. Bring your seeds and seedlings to swap (free packet of seed to start you off)

New to seed collecting? Take this opportunity to learn. Several local stalls. Refreshments.

Donations

Transition Falmouth

Engaging the community in designing a vibrant low energy future for Falmouth and the surrounding area.

Web: www.transitionfalmouth.org.uk   Tel: 01326 317 587

Transition Falmouth

Engaging the community in designing a vibrant low energy future for Falmouth and the surrounding area to face the twin challenges posed by Peak Oil and Climate Change.

Please join us and participate in the process.

info@transitionfalmouth.org.uk
www.transitionfalmouth.org.uk
c/o Lorely: 01326 317 587

TRANSITION TOWN BRIXTON
presents
Energy beyond Oil

An illustrated talk by Paul Mobbs
Thursday 17 May, 7.30, Lambeth Town Hall

A clear look at how much is left and how our energy use may change in the future.

visit www.transitiontownbrixton.org for more events
Transition Town Brixton, the community led initiative to design Brixton's pathway to a better low energy future.

Transition Town Totnes Traffic and Transport Group in association with **Schumacher College** jointly present;

Redesigning Our Towns and Cities for Life Beyond The Car.

With **Herbert Girardet**, author of 'Creating Sustainable Cities' and **Peter Lipman** of Sustrans.

SHARE THE ROAD

Freemasons Hall, Totnes.
Wednesday 12th September. 8pm. £4 (£3 concs).

Given the dependence that the economy of Totnes has on car-bound visitors, and given, in an era of rising fuel prices and our increasing oil vulnerability, the time is right to look at the question, what will we do when there are less cars on the road? How will we get around? How will others visit us? How can we ensure mobility into the future? Earlier this year, the building work in the centre of town temporarily closed the town's main car park, and the impact on local businesses and traders was keenly felt. These two events will explore these key questions.

Imagining Totnes Beyond the Car?
An Open Space Day.

Bring your idea, passions and enthusiasm. Lunch, tea and coffee will be available.

Methodist Hall Totnes.
Saturday 15th September. 10am – 2pm. Free.

www.transitiontowns.org/Totnes
TOTNES

Transition Penwith presents:

PEAK OIL
LOCAL SOLUTIONS TO A GLOBAL CHALLENGE
with Richard Heinberg

Richard Heinberg, a world leading authority on Peak Oil, will give a presentation on the most recent evidence regarding the global oil production peak, its likely consequences, and what we can do at the local level to address the impacts.

Keynote Address: MP Andrew George
Friday 24th November, 7.30pm
St. Johns Hall, Penzance
Cost: £5.00

A limited number of subsidised places available
Advance Booking Recommended!
tel: 01736 793876
email: peak_oil@virgin.net
www.transitionpenwith.com

www.richardheinberg.com   www.oildepletionprotocol.org   www.peakoil.net

This event marks the launch of a new partnership initiative, Transition Penwith, which seeks to address our transition from an oil based culture to one based on sustainable, ecological design principles. Supported by:

Friends of the Earth   coast   ecodrive

Transition Town Totnes presents;

zerocarbonbritain
an alternative energy strategy

A Talk by **Paul Allen** (Director CAT) & **Tim Helwig-Larsen**, co-authors of the report.

Paul Allen and Tim Helwig-Larsen talk about the Zero Carbon Britain report, which sets out a plan for how Britain can become carbon neutral over the next 20 years, through a mixture of innovative energy conservation methods and massive renewable energy generation. The talk is a great opportunity to gain insight and clarity into how we can go about achieving our vision for lower energy consumption in the near future, exploring how the Z.C.B Strategy can be applied locally here in Totnes. As the town starts work on its Energy Descent Plan, this event provides timely and vital ideas for anyone interested in energy use, carbon emissions, and reducing our environmental footprint.

Wednesday October 31st. 8pm.
St Johns Church, Bridgetown, Totnes.

£4 (£3 concs).
Totnes Pounds accepted.
www.transitiontowns.org/Totnes   TOTNES

ple to talk with each other and feel some degree of support in exploring these issues.

Although your awareness-raising process is, on the surface, about informing people and disseminating ideas, it is also, perhaps more importantly, about getting people talking to each other, starting to build the social networks on which your Transition Initiative will depend. Make sure any event gives people the time to talk to the person next to them.

You might also run an evening class, go into schools, write articles for the local paper, or get something on the local television. There is really no clear way of knowing when this stage has been done sufficiently to allow you to move on to Step 3 – you just have to gauge that yourself somehow. I was only able to effectively assess the impact of what we had done when Richard Heinberg was in Totnes in December 2006 and at the beginning of his talk he asked the audience how many of them were familiar with the concept of peak oil. Three quarters of the room (which contained about 350 people) put up their hands.

This stage also allows you, if you are new to the town you are working in, to meet people, to see who are the people who come to all these events, who may become your key allies. All this will stand you in very good stead when you kick off the process proper. The social network building aspect of this is as important, if not more so, than how many people are able to tell you who M. King Hubbert was and what the annual oil output of Mexico is.

Transition Tip
- *One way you can use your film screenings to draw in official bodies is to invite a member of the local authority, preferably one who makes decisions on energy and environmental issues, or perhaps a planner, to be a member of a panel to comment on the issues raised by the film. This will be beneficial in two directions: it introduces them to the issues your Initiative is exploring and also to the organisation, but it also allows you to question them about their thinking on such issues.*

3. Lay the foundations

It is extremely unlikely that you will be starting a Transition Initiative in a place where absolutely no environmental initiatives have ever happened before. It is possible that such places exist: if you are in such a place it might be worth contemplating why nothing has ever happened there before! Within the community there will be people who are just finding out about environmental ideas, people who have been familiar with the intellectual side of it for years but haven't done much practical action, those who are gardeners, growers and builders, and people who are burnt out from doing similar stuff for years while no one listened.

There is also a range of official and semi-official organisations and bodies, from local government to Women's Institutes. It is essential at this stage that you network with these groups, and make it clear that this is a process of supporting and collaborating with them, rather than duplicating their endeavours or, worse still, dismissing their years of hard work as somehow irrelevant. Offer presentations to all the existing environmental and decision-making organisations in the town.

When introducing your Initiative to other groups, give a concise and accessible overview of peak oil, what it means, how it relates to climate change (this may be an important point with some green groups who are committed to tackling climate change but not really *au fait* with peak oil and the relationship between the two), how it might affect the community in

Schumacher College, Transition Town Totnes, South Hams FoE and Totnes Renewable Energy Supply Company (TRESCO) jointly present;

The "Local Responses to Climate Change" Series.

Wednesday February 7th. 8pm.
Contraction and Convergence.

A talk by Aubrey Meyer
Originator of the concept.

Aubrey Meyer co-founded the Global Commons Institute and has received the Andrew Lees Memorial Award in recognition of his proposition of 'Contraction and Convergence' (C&C) to combat climate change. The New Statesman described Aubrey as "one of the ten people in the world likely to change it". He is also a professional violinist.

St. John's Church, Bridgetown. £4 (£3 concs).

Wednesday February 21st. 8pm.
Community Responses to Climate Change.

A Talk by Tony Juniper,
director of Friends of the Earth.

Tony Juniper is Executive Director of Friends of the Earth. He has been closely involved with climate change campaigning since the Rio de Janeiro Earth Summit in 1992. He was with a team of campaigners helping to nurture the Kyoto Protocol into existence and has worked to influence many aspects of UK policy on climate change.

St. John's Church, Bridgetown. £4 (£3 concs).

Schumacher COLLEGE   **Friends of the Earth**

question, and the key challenges, as well as the key opportunities it presents. Set out your thinking about how the Transition process could act as a catalyst for getting the community to explore this and to begin thinking about grass-roots mitigation strategies.

You do need to be a bit careful in jointly organising events with other groups: when it works it's great, but if it runs into problems it can be difficult to keep everyone happy. You will need to ensure that each group is happy with how the event is presented, promoted and facilitated. For example, we have organised some talks with Schumacher College and the local FOE group (see poster above), which were very successful. Part of this phase also involves reaching out to groups who are usually bypassed or ignored by environmental groups, for example the local Chamber of Commerce or

the Conservative Association. If this is going to work it will need the input of a broader range of bodies than has been the case in the past.

In essence, 'Laying the foundations' is about networking with existing groups and activists, and stressing that this Transition Initiative is not a process of duplicating their work but of requesting their input in a new way of looking at the future. Acknowledge and honour the work they do, and stress that they have a vital role to play.

Transition Tip
Invite other local groups to co-present events with you, and design as many events into your programmes that involve other groups as possible.

4. Organise a Great Unleashing
I use the term 'Unleashing' because that is the sense that this event should embody. Through the first three stages, ideally you now have a groundswell of people fired up about peak oil and climate change and eager to start doing something. The aim of this event is to generate a momentum which will propel your Initiative forward for the next period of its work.

The Official Unleashing of Transition Town Totnes was held in September 2006, and had been preceded by about ten months of talks, film screenings and so on. By the time of the Unleashing, we felt that there was sufficient energy in the town to do this successfully. This was based, entirely unsubjectively, on the fact that numbers attending events were steadily increasing, more people wanted to stop us in the street to talk about it, and also the fact that we were getting impatient to kick it all off. How you judge when to do your Unleashing is entirely a matter of collective judgement.

Some groups, such as Transition Penwith, started pretty much from cold with an

Tools for Transition No. 7: Making the most of your public events

A film screening is much more than just an opportunity to sit a load of people in front of a screen. Likewise, a talk is more than just the chance to hear the musings of a well-known thinker on a particular subject. Both are opportunities to get people talking to each other, networking, building social connections. Indeed, one might argue that these are far more important than the film itself; they could, after all, just borrow the DVD and stay at home. It is also important that you work into these events what we might call "digestion time", that is, time to chew over what people have heard, rather than just dumping information on them and then sending them out, blinking and bewildered, into the world. Here are some of the things you might expect at the average Transition Initiative film screening or talk:

'Think and listen'

Usually just before the film or talk, and immediately after, we invite people in the audience to turn to the person next to them and to spend five minutes each way, talking and then offering listening to the other person. Before the film the theme is "who are you, what has brought you here, what are your hopes for the evening?", and then after the film or talk to discuss their thoughts about that. Before the film it is a very useful tool for starting to build a buzz at the event, people really enjoy meeting other people and feeling, from the outset, part of a larger body and an energetic movement.

After the film it is a very good way of helping people to order their thoughts in advance of being able to ask questions and/or contribute to a group discussion. It is also a good opportunity to 'digest' the information in the film, which may have been unsettling or even distressing for some.

The Comments Wall

This is another tool for helping people to feed back their thoughts on the evening, and to be able to voice thoughts or opinions they felt unable to contribute by standing up with a question or a comment. On the back wall, where people leave the venue, we stick together lots of sheets of flip chart paper, stick them to the wall and provide lots of pens so that people can write their comments. This can yield very useful feedback sometimes, and is a valuable opportunity for the shyer attendees to have their say.

Celebration!

Celebrating is something that we earnest greenies aren't that good at, but it is a key element to build into each stage of our work. When Transition Stroud showed *The Power of Community* a while ago, they followed the

Attentive audience at Transition Bristol's BIG Event, November 2007.

screening with food, drink and live music. While it may prove impractical that every event features a celebration on such a scale, it is a powerful concept to bear in mind and to work in where possible.

Harvesting email addresses

These events are your opportunities to start building up a database of email addresses of supporters in the community, which you can then use for sending out a regular bulletin, updates or news of particular events. With TTT, on the door of each event we have a list and we invite everyone as they arrive, to leave their email addresses, stressing the confidentiality and that they are not given out to anyone else. This list becomes an invaluable resource. Always be mindful of issues around data protection, and always send any group emails as BCC (blind carbon copy).

The Post-It Note tool

This doesn't have a snappy name, but it is a great tool. It originally came from something similar they did when they showed *The End of*

An early post-it note experiment, Totnes, December 2005.

Suburbia in Machynlleth in Wales. We adapted it slightly to suit Totnesian ends. When people arrived for the film, they were given four different-coloured post-it notes. They were asked to write on them accordingly:

Pink – One thing I can do
Yellow – One thing Totnes can do
Orange – One thing the Government can do
Green – One other thought

When giving attendees their post-it notes, don't tell them what they are for, so as to build a sense of anticipation.

A follow-up

This works particularly well after the post-it notes exercise, but you may like to adapt it for anything that involves the gathering of people's ideas. After the event, type up the post-its and then email them out to everyone who attended. This is a very useful tool for keeping the ideas current in people's minds, as well as for remaking the point that their thoughts are part of a powerful surge of thinking that is being unleashed within the settlement in question.

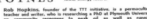

Transition Town Totnes
presents:

The Official Unleashing of Transition Town Totnes.
Totnes Civic Hall. 8pm.
Wednesday September 6th.

Transition Town Totnes is a new initiative seeking to engage the community in the process of designing a practical pathway to a more sustainable society. This evening will be opened by Pruw Boswell, the Mayor of Totnes, and will feature presentations by:

Dr. Chris Johnstone, author of *'Find Your Power'*, whose work specialises in addictions and ecopsychology, will talk about how a community such as Totnes can find its collective power and strength to embark on this great adventure to a lower energy future.

TRANSITION TOWN
TOTNES

Rob Hopkins, founder of the TTT initiative, is a permaculture teacher and writer, who is researching a PhD at Plymouth University on community responses to peak oil as well as running www.transitionculture.org, a website which explores these issues.

The evening will be inspiring and will set out how the TTT initiative will work and will also invite your thoughts on the transition to a low energy, more localised Totnes. All are welcome. Free of charge.

For more information call 07868604454
or email: robjhopkins@gmail.com
www.transitiontowns.org/Totnes
www.transitionculture.org

PDF created with pdfFactory Pro trial version www.pdffactory.com

Photos credits: Images grabbed from video footage by Sally Hewitt

Unleashing, because they had the chance of having Richard Heinberg present it, a rare enough opportunity. The ideal though, as I see it, is like one of those toy volcanoes that children like: you gradually add a bit of vinegar, a bit of baking powder, a bit more vinegar, a bit more baking powder, until the pressure inside builds to an unbearable point, and then BAM, you hold your Unleashing. It marks the arrival of the project, and it is a celebration of the community's desire to act.

It is important to stress, as Chris Johnstone did at the TTT Unleashing, that this is an historic evening, the beginning of the great change, the evening people will look back to as the evening where it all started. There is a balance between it being perceived as too flaky for the serious environmentalists, and too dry for those who like to be more emotional about things. It is a balance Chris strikes perfectly, and I'm sure others can too. His angle is that, as observed in his work with addictions, once we decide to act, we find our power, and that in seemingly impossible situations, it is by doing it that we find qualities and strengths we never knew we had.

Other things we did at the Totnes Unleashing were getting people talking to each other in pairs about their concerns and fears in relation to peak oil and climate change, as well as their visions for the future. They were invited to write these up on post-it notes and put them on the wall. These were subsequently typed up and emailed out to everyone who attended. As many opportunities as possible for people to meet each other and to talk were built in. It should be a memorable and historic occasion. How you do that will be different in each community.

An Unleashing will inevitably be different from the usual peak oil presentation. It is not about bad news, gloom and doom and passing on to people the information about just how

precarious our situation is. It is a celebration of the possibilities that lie ahead of us if we act together with imagination, having harnessed our collective genius. It is a celebration of the community's resourcefulness and creativity. Certainly the Totnes one created a huge amount of energy and goodwill that has driven it forward ever since. This also happened in Penwith and Lewes. One thing we should have done and which I would recommend is to draw up a list of people to invite, councillors, planners, politicians, local movers and shakers.

An Unleashing is not something to be organised lightly. It is a one-off opportunity to bring all those people together and to launch the Transition Initiative. If you get it right, people's lasting impressions will be that this is a dynamic project which is going to do great things. A poorly organised, ill-attended, half-

Transition Culture   April 26th 2007

LEWES NOW OFFICIALLY UNLEASHED!

On Tuesday night Transition Town Lewes held their Official Unleashing at Lewes Town Hall. Only the third town to organise an Unleashing, Lewes put on a wonderful event, one which will have been very memorable for all those involved. Over 400 people packed into the Town Hall to hear John Webber, the chair of TTL welcome them and introduce the evening. I then spoke for 45 minutes about peak oil and energy descent, where the Transition process emerged from and how it is being applied in various places.

Then Adrienne Campbell from TTL spoke about what they had done so far, and the events that are coming up in their excellent new programme. She also spoke very movingly of how in 1900, all the businesses in Lewes were owned by local people, listing some of them and what happened to them. She invited everyone assembled to come to their first Open Space day, on food, which is coming up soon.

Then Dr Chris Johnstone spoke about finding your power through engaging in energy descent work. He explained the Stages of Change model that emerges from the addictions field, and how it can be applied to Transition work. He talked about 'dreamblockers', the voices in our heads that stop us from doing things, and how in driving forward the energy descent planning process we will face these dreamblockers within our communities, those who tell us it can't be done.

He finished his talk with a rousing salsa sing along (you had to be there!). Finally the Mayor of Lewes spoke, and offered the unanimous support of Lewes Town Council. There was a great sense of enthusiasm and passion for the project, and people began to sign up for leading some of the subgroups. In the hall outside were a number of stalls and stands from local groups and people stayed for a long time discussing the evening and the issues it raised, oiled by local wine and fruit juice.

Transition Town Lewes is a very dynamic group, the first Transition Initiative in the UK to have its own T-shirts (a few were hand-made for the occasion!) as well as its own office! Congratulations to everyone involved for such a successful evening, as well as for all that they have done so far.

The talk I gave at this event can now be seen on YouTube, at: www.youtube.com/watch?v=Vb7MncibAHs

hearted Unleashing will make the next phase of your work an uphill struggle. I would think that six months to a year after your first film screening is about right, but clearly that depends on your situation. It should be a powerful, passionate, informative, and inspirational evening that people will remember for many years to come. Don't rush it.

Transition Tip

Your Unleashing need not just feature talks. It could include music, performance, images from local history. It is also useful to create space within it for local groups to present themselves. At the Unleashing of Transition Town Lewes, adjoining the main room was a room where all the various local 'green' groups and community organisations had stalls.

5. Form groups

Part of the process of developing an Energy Descent Action Plan is that of tapping into the collective genius of the community. One of the most effective ways to do this is to set up a number of smaller groups to focus on specific aspects of the process. Each of these groups will develop their own ways of working and their own activities, but will fall under the umbrella of the project as a whole.

Transition Tip

Each meeting of a working group could invite a 'witness', someone who has expertise in the field in question, who could then be asked for their perspective on the issues, their experience and their ideas for subsequent 'witnesses'.

As organisers of the initiative, you can be quite proactive in getting these groups to form. In Totnes what we did was to design the programme so as to encourage specific groups. For example,

to get the Food group going, we first ran an evening event called 'Feeding Totnes: past, present and future', where speakers addressed each aspect of the issue in turn. This drew in many of the people in the town with an interest in food.

This was followed three days later by an Open Space Day on food. This explored in depth the possibilities of the relocalisation of food in the Totnes area. From this meeting a number of initiatives emerged, and people came forward to run the Food group. We have since used this model to start a few of the other groups.

In TTT, we have set out a collection of guidelines that we ask those considering forming new groups to read through first. They run as follows:

- Each group should have a core of people who steer it, and who meet regularly, but also be open to whoever else wants to come.

- Each group should be continually asking itself 'Who isn't here that should be here?', that is, always being open to exploring new 'avenues' by which new people with relevant skills can be drawn into the group.

- The key task of each group is to explore the question: 'What is a vision for a low-energy Totnes in relation to this field, and what might a timetable for that look like?' The group is assembling ideas and information that will enable them to put together their section of the Totnes Energy Descent Action Plan.

- Each group will have access to the relevant section of the project's website, and will also be able to use the logo in its publicity materials. In exchange, it will keep a record of its business on the website, be it as minutes or as notes. At present the Transition Network offers Wiki webspace to

Tools for Transition No. 8: Writing a good press release

Dealing with the media will rapidly become a key element of the work of your Transition Initiative. One of the first skills you will need to perfect is the writing of good press releases. There is no avoiding the writing of press releases, they are key to communicating what you are doing to the outside world. The first thing to remember is that journalists are extremely busy people. You need to catch their attention in the header and first line of the release, be as concise as you can, and give them what they need in a format that is as easy for them to use as possible.

When you write your press release, it is important to consider exactly what it is that you want to say. What do you want people to get from reading it? What is unique about the story you want the press to give their precious column inches to?

You also need to think about timing. When do you want it in the publication in question? Daily papers clearly have a shorter lead-in time than quarterly magazines. Check the deadlines for copy and make a schedule for your releases based on your programme of events. It is also worth bearing in mind that they often start setting out the main features of the publication before their official deadline date, so getting it to them earlier increases the likelihood of a better spot. Make sure you identify the best person to send it to and address it specifically to that person. It is worth following up your email with a hard copy if you have time.

A good formula to use when constructing your release is 'AIDA':

A Get the reader's **Attention** (with good headline, sub-head and first sentence)

I Give the **Information**

D Give the **Details** (when, where, contact info)

A Inspire the reader to **Action** (with an enticing last sentence)

Structuring a press release
I suggest structuring a press release as follows:

Heading
"Press Release", in large, bold letters.

Date
Is this press release 'For Immediate Release' or to be 'Embargoed Until' a specific date when you want it to appear?

Contact details
Include the name, address, phone number, email address and website of the body sending the release.

Headline
This needs to be the catchy headline that draws them into the piece. You'll get a hang of the kind of thing by reading through the kinds of publication you are sending your press release to. The likelihood is they'll change it, but it is your first opportunity to grab their attention.

Subhead

Encapsulate your release in one or two intriguing short sentences if possible.

Opening paragraph

This sums up the content of your press release and makes people want to read it. It needs to contain what is newsworthy about the rest of the text, and give the reader a summary of the story. The first 20-30 words are the key to engaging the reader. You will need to give people a short (5-6 word) reminder of what your project is about. Our local paper, the *Totnes Times*, always describes TTT as "the project exploring the town's transition to life after oil".

The rest

It is best to try to keep the rest of the text on one side of A4 paper. This saves second pages getting lost and is more concise for busy editors. Illustrate your text with quotes from members of the project. Here is an example from a TTT press release to announce the release of the project's Autumn 2007 programme, " 'We think this is our strongest programme yet and that it is a testament to the energy and dedication the community has put behind this initiative', said Rob Hopkins of TTT."

Pictures

You may want your story to be illustrated by a picture, in which case attach a high resolution jpg file of the image to your email or note at the bottom of your release that you have pictures if wanted. Make sure it is a good quality, well composed picture, and try to imagine it reproduced in black and white in a newspaper; images with high contrasts of light and dark work better. Sending a picture can sometimes make the difference between the paper using your press release and not.

At the end

At the end put any further useful information, such as where tickets for the event can be bought, or contacts for arranging interviews with the visiting speaker if appropriate. In terms of getting your press release out to as many people as possible, it is useful to have a good database of media contacts. Keep this up to date, and where possible, visit the relevant editors so you have a good working relationship with them. It is also useful to follow up your press release with a phone call to see that they have received it and whether they need any more information – your phone call can often make the difference between your story being covered or not, provided you time it right – catching an editor just as he is trying to get his edition to press can be counter-productive. Try to avoid sending your press release as an email attachment, or if you do, also include it in the body of the email.

You can see a sample Press Release in Appendix 6, p.223.

new Transition Initiatives, which is easy to maintain and update.

- Each group will also be able to use the Wiki website, allowing it to post initial drafts of documents or of its section of the Energy Descent Plan, in such a way that others can edit it online – a powerful collaborative information-building tool.

The final thing to say on the question of setting up groups is that you cannot assume that everyone who offers to form and facilitate a group actually has the skills to do so. It became clear after a while that it was useful to offer training in facilitation and designing successful meetings to all our group convenors. We organised a day with Andy Langford and Liora Adler of Gaia University on the subject of Designing Productive Meetings. This covered basic tools like 'Go-rounds', 'Think and Listens' and so on (see Tools for Transition No. 9, pp,164-5), and was extremely useful.

It is also useful for these groups to meet each other on a monthly basis (at least). In TTT, the facilitators of each group meet on a monthly basis. Each group sends one person to that meeting. We meet and then have lunch together. Designing in as many opportunities for the groups to network and meet each other as possible is crucial.

Transition Tip
It may not always be necessary to actually start a new working group. Sometimes there may be existing groups in the area who have done lots of work on a particular subject. There may, for example, be strong renewable energy groups, or local food groups. Consider avoiding duplication by going to them and asking if they may like to take on the role of being a Transition group, and feed their ideas into the EDAP process.

6. Use Open Space

Open Space Technology is an extraordinary tool. It has been described as "a simple way to run productive meetings, for five to 2,000+ people, and a powerful way to lead any kind of organisation, in everyday practice and ongoing change". In theory it ought not to work. A large group of people comes together to explore a particular topic or issue, with no agenda, no timetable, no obvious co-ordinator and no minute-takers. Yet by the end of the process, everyone has said what they needed to, extensive notes have been taken and typed up, lots of networking has taken place, and a huge number of ideas have been identified and visions set out. (See page 168 for how to run an Open Space event).

Transition Tip
Make sure that you have given a lot of consideration to the focusing question that will underpin your Open Space Day. It should be one that is relevant to those attending and that will draw in those people passionate about the subject, for example "How might our town feed itself beyond cheap oil?", or "What is the role of education in a lower-energy future?"

At TTT Open Space Days, the ideas generated in the event are uploaded in real time, live onto our website. So far these have been on Food, Energy, Housing, Economics, the Arts, the Psychology of Change, Education and Transport. This requires someone prepared to 'scribe' the notes that emerge from the different groups, access to broadband, two laptops, a memory stick or writeable CD, and someone able to upload onto the website. A digital camera is also useful. The beauty of posting it live onto a Wiki site is that anyone who is following

The opening circle of TTT's first Open Space day on food, October 2006.

Transition Tip

Sustaining energy levels can be done in different ways. Ensure that tea and coffee are available throughout the day. If you can organise lunch for people it helps a great deal, as it gets round the problem of people drifting off for lunch and then getting back late. One thing we did at one Totnes Open Space day was have a group of musicians who were playing later that evening in Totnes come and play to the Open Spacers at lunchtime. It was good advance publicity for them, and it completely changed the focus and re-energised the day.

the event anywhere in the world can send in their thoughts on the subject. It also means that at the end of the day you don't have loads of notes that some poor soul has to take home to type up, but that all the outcomes of the day are typed and available for people to muse over and add comments to when they get home.

Some time in advance of each Open Space, draw up a list of the key people you feel should be there for that topic. Send them a personal, not a generic, invitation. Make them feel that you have specifically invited them because of their expertise on the subject. There are other tools which are similar to Open Space, such as World Café (see pp.184-5) which have many of the same outcomes. The essence is to get people talking, building relationships, discussing ideas and making connections. It can do a great deal to identify priorities for the work ahead in relation to that subject.

7. Develop visible practical manifestations of the project

It is easy to come up with ideas, but harder to get practical things happening on the ground. It is essential that you avoid any sense that your project is just a talking shop where people sit around and draw up endless wish lists. Your project needs, from an early stage, to begin to create practical manifestations in the town, high visibility signals that it means business. The power that doing this has to affect both people's perceptions of the project and their willingness to engage is huge.

These manifestations can take a variety of forms. It might be productive tree plantings, solar panels, or hemp/lime plastering. It could be a beautiful cob bus-shelter or an alternative currency used for a defined period. They should, at this point, be both uncontroversial and photogenic.

Tools for Transition No. 9: Designing productive meetings

Very soon you will find that you are having meetings, and that you need to get a lot done in a limited amount of time. For most of us, the idea of meetings, indeed the very mention of the word, leads to a sinking feeling, akin to that associated with the words "going to the dentist" and "doing your accounts". One notable sustainability project that I followed for a while in the UK in the early 90s, folded after four years of innovative and productive work. I asked one of the founder members why. "I think we just met each other to death", he told me. It doesn't have to be like this. There are a number of tools that we use that make our meetings far more productive, and, dare I say, enjoyable.

'Go-Rounds'

These are key in a meeting, and in TTT we usually use them in two ways. Firstly at the beginning of a meeting we do a go-round of what has been happening in each person's group since the last meeting. We give each person 5-10 minutes, and avoid interrupting them or getting into a discussion about what they have said. During their time they are also asked to begin with "how am I feeling right now", and then, once they have updated the meeting on progress in their group, to highlight what they would like to be put on the agenda for the meeting.

In the group's early meetings, when the people are still getting to know each other, we also put in a "throwaway question" for the end of each person's time. These can be things like "the best meal I ever had", "the most beautiful place I have ever been", or "the greatest piece of music I ever heard". There are of course many variations on this, but they enable the group to get to know each other a bit better, and to relax more with each other.

Open Agendas

We don't prepare an agenda in advance. The danger with prepared agendas is that there is a sense that somewhere 'they' have set the agenda for the meeting and have decided in advance what will be discussed and even what will be decided. It is centrally important that any sense of this kind of hijacking of the group's intentions is overcome. We start the meeting with a blank sheet of flip-chart paper, and as we do the introductory go-round, we write up the issues and questions that people want to see put on the agenda. Once this list is complete, we collectively go through the list and label each item between one and three, one being "this must be discussed today", two being "ideally we'd like to talk about it today, but at a push it can wait", and three being "this could wait until next time".

We then assess how much time we have left in the meeting and allocate each item a time limit, which we then try to stick to religiously. It is a good idea to also timetable the rest of the meeting so that at the end you close with a non-contentious issue, so that you don't all leave seething with each other! It is important to keep

Running a productive meeting is a learnable skill. Here TTT members are doing a training day with Andy Langford and Liora Adler of Gaia University.

the agenda visible to all, and to check with everyone that the item has been satisfactorily dealt with before it can be checked off the list.

'Think and Listen'

These have been explained in detail above, but they can also be useful in meetings when you want to take a short break from addressing a big issue, and give people the space to order their thoughts in advance of tackling it. It is also a useful tool for keeping people's brains alive — halfway through a meeting eyelids can occasionally start to become heavy, and a 'Think and Listen' can be quite an energising and refocusing tool to use at that point.

Clear Beginnings and Endings

Make sure the meeting starts with something to mark its opening, perhaps a minute of silent reflection, or even just announcing that the meeting has begun. At the end, it is good to have something that formally closes the meeting, so that it feels like it has a formal sense of closure, rather than just drifting off into the next thing.

Celebration!

Again, this is a key component of your meetings. One of the easiest and most satisfying ways of doing this is to eat together. At TTT meetings, we all bring a dish to share and end our meetings by having lunch together. As well as satisfying the need for food, it also addresses a longer-term need for getting to know each other better in a non-meeting context.

The Layout of a Typical Transition Town Totnes Project Support Group meeting

- Welcome

- Go-round in which each person has five minutes, uninterrupted, to talk about everything that is happening in the group that they facilitate.

- We then brainstorm what people would like to be on the Agenda for the meeting.

- Go through that list and prioritise items, dismiss those that can wait until the next meeting, deal with those that require simple answer or brief agreement, then allocate time to the discussion of each remaining subject.

- At the end, arrange date for the next meeting, then eat together for the last hour, to discuss any other issues and to build our 'social glue'.

*Lewes began with forming a steering
group but one still discussing what
its demise might look like. They
began their Awareness-raising stage
and then planned an Unleashing.
They then decided that they hadn't
done enough awareness-raising, and
postponed the Unleashing.*

*They showed films and held talks,
as well as working with various local
groups (Laying the Foundations, Step
3). Their Official Unleashing on April
24th was followed by their forming
sub-groups (Step 5) and hosting their
first Open Space Days (Step 6). They
have followed a pretty logical
sequence through the Twelve Steps.*

In permaculture, we talk about the need to observe a plot of land for the first year before making any interventions or completing any design. The same goes for a Transition Initiative. Your first year is a time for networking, brainstorming, awareness-raising, information gathering. It is the time when you are gathering the pieces that are later assembled in an Energy Descent Action Plan. You don't want to start doing projects which, once you have completed your Energy Descent Action Plan, you realise are in the wrong place and not actually properly thought out. While it is essential to take your time and plan properly, there is a balance to be struck here; you also need to carry the community along with you. Small highly visible projects will allow people to see that you mean business, that you are here to stay, and will give them a tangible sense of what you are talking about.

People need to get a sense of the whole, and to see things happening that they can go home and tell their friends about. These practical manifestations will also bring into the project the people who have spent the first few months sitting back, saying, "We'll see. I've heard all this before, this is just another of those flash-in-the-pan projects, I'll keep an eye on it." When they start to see infrastructure going in, it becomes infectious, they want to be a part of it.

If you have done the previous steps well, you may well find that the sub-groups start developing their own practical projects automatically. As the momentum builds, you will find practical manifestations bursting out all over the place.

Another spin-off benefit of these practical projects is their great team-building potential. A group meeting regularly to discuss food issues is one thing. If that group meets and plants an orchard in a day, shares a picnic and leaves with a sense of great achievement, that is very powerful in terms of building the dynamic of that group. A Transition Initiative with dirt under its fingernails will carry a lot more credibility. In Totnes, the Totnes Pound has been our most successful manifestation of this Step.

Transition Tip
Make sure that you get good publicity for all these

Photo by Nevenka Mulej

The Totnes Mayor and Town Crier with Noni McKenzie of TTT and a very large carrot launch the second Totnes Pound and the TTT Local Food Directory on the wettest day in living memory.

practical manifestations. This will be very helpful in building a widespread confidence in what you are doing. Involve local schools, local dignitaries, and design events that grab the attention, be they tree plantings, natural building projects in schools or delivering books to the local library. Make them as photogenic as possible.

8. Facilitate the Great Reskilling

I believe that one of the main factors contributing to the sense of panic that often sets in immediately after an awareness of peak oil, especially for young men, is the realisation that we no longer have many of the basic skills our grandparents took for granted. One of the most useful things a Transition Initiative can do is to offer widely available training in a range of these skills.

What skills ought we teach? We need to enlarge our ideas of what these might be. Some research is useful, in terms of what skills people used to have that might still be appropriate, as well as looking at the skills people have or need now. These Reskilling events fulfil a few different functions:

- They bring people together, relaxing and learning new skills
- They build networks
- They build a fundamental sense of 'can do'
- They can create a link between old and young, as skills are passed on
- They can be practical events which actually put something in place, like a natural building day that produces a cob bus stop or similar, thereby offering an opportunity for creating Practical Manifestations (see Step 7, p.163)

Work with existing groups, local sustainability centres, colleges and so on where possible. Draw on local skills wherever you can. It is

Transition Falmouth's Darn It group provide re-skilling in a range of crafts.

Learning by doing: Kinsale students learning cob/cordwood building through hands-on experience.

great if you can design these events in such a way that students from the first running of a course can help teach the students the second time it runs.

To begin with, your Great Reskilling will largely consist of one- to two-day courses, or longer evening classes like 'Skilling Up for Powerdown' (see page 194). In time you might be able to offer something like the two-year full-time Practical Sustainability course at Kinsale FEC in Ireland. On that scale there is a lot you can do in terms of outreach and engagement.

A Transition Initiative invites a community to undertake a journey, to embark on a collective adventure. Very often in modern society people feel disempowered, such that even changing an incandescent light bulb for a low-energy one is too much effort. Your Great Reskilling should give people a sense of the power of solving problems, of practically doing things rather than just talking about them, and of the sense of belonging that comes from working alongside other people. Above all it should be fun.

APPLYING THE TWELVE STEPS No. 2: KINSALE

Kinsale took an odd way through the Twelve Steps. In effect, it jumped straight from zero to having an Energy Descent Plan. Although there was a small amount of Awareness-raising (2) and Open Space Events (6), most of the other stages, such as forming the groups (5), happened within the College, which already had the practical manifestation (7) and the Great Reskilling (8) pretty well established. Because for the larger community of Kinsale, steps 1-11 are virtually non-existent, Transition Town Kinsale is basically starting from scratch from number 1 (for more, see p.122).

"We learn to do something by doing it. There is no other way."

– John Holt

Tools for Transition No. 10: How to run an Open Space Event

If you are a control freak, you will hate organising an Open Space event! It involves a lot of trust that the process will work but at the same time I have never seen or heard of one not working. Open Space[1] is a powerful tool for engaging large groups of people in discussions to explore particular questions or issues. It can be used with groups from anything between 10 and 1,000 people. Open Space has Four Rules and One Law (the Law of Two Feet).

The Four Rules state:

1. Whoever come are the right people.
2. Whatever happens is the only thing that could have.
3. Whenever it starts is the right time.
4. When it's over, it's over.

and *The Law of Two Feet* states that:

"If, during the course of the gathering, any person finds themselves in a situation where they are neither learning nor contributing, they must use their feet and go to some more productive place."

Key to a successful Open Space event is the question – which is usually in the title of the event – which sets the ground for what is to be under discussion that day. Some examples from Totnes include:

- How Will Totnes Feed Itself Beyond the Age of Cheap Oil?
- Powering Totnes Beyond Cheap Oil . . .
- The Economic Revival of Totnes – how can we build a sustainable, equitable and healthy economy in Totnes?

You may choose to invite specific people, or just leave it open to whoever turns up. Your venue needs to be large enough to take those who attend sitting in a large circle. You need walls on which you can stick things, and you need enough other space for several break-out areas.

When people arrive they take a seat in the circle, and when everyone is gathered the event begins. In the centre of the circle is a pile of sheets of A4 paper and pens, and on the wall is

	1 (Table by the window)	2 (The one with the flowers on)	3 (Side room)	4 (Room upstairs)	5 (Dining room)
10am - 11.30					
11.30am - 1pm					
2pm - 3.30pm					

Figure 19. A model example of an Open Space open agenda.

an empty timetable, with the timings of the different sessions on one axis, and the various breakout spaces on the other (see foot of opposite page).

Each of the squares is the size of an A4 sheet. Explain to people the rules of Open Space and that the only prerequisite for proposing a question is that you will host that discussion and scribe the conversation for the benefit of anyone unable to participate, and write that name on the sheet, and post it on the timetable. Then say "Go!". This is the nerve-racking bit: your heart will be in your mouth the first time you do this, worrying that no one will come forward, but then one person does, and the floodgates open.

What follows is a ten-minute mêlée of people proposing questions and sticking them up. You may well end up with more questions than you have slots available, in which case consolidate some of them together, for example, three or four may be on different aspects of community energy generation and so could be rolled up into one session on that subject. Once your timetable/agenda is full, allow people a few minutes to look at it and work out what they want to go to, and then ring a bell, or something similar, to announce the convening of the first session.

In theory, the rest of the day will organise itself. Each break-out space should have plenty of flip-chart paper and pens. At the end of each session, ring a bell to let people know it is finished, then go round and collect up the note-filled sheets, and put them up on the wall in the area you have pre-designated as the 'Market Place'. You may also choose to have someone

typing up the sheets, if you are posting the proceedings live on the web, or even if you just want to get to the end of the whole thing without endless typing to do. A bit like washing up, typing up a pile of flip-chart paper looks far more daunting the next day.

Tell people when each session starts, and remind them about the Law of Two Feet. The event will then, basically, run itself. Leave 30-40 minutes or so at the end for a go-round, for reflections on the event and the process itself, rather than issues raised. Harrison Owen's book *Open Space Technology*[2] sets out ways that Open Space can be used in longer two- to three-day events, which people leave with a full set of minutes, but this scale of event is more appropriate for corporations than for Transition Initiatives, as it would be difficult to find many members of the community able to give up 3 days of their time.

Open Space is surprisingly easy to run, and an amazingly powerful way of exploring issues. What it does is draw out all those who are really passionate about a subject. For your first one you might find it useful to have someone with prior experience of running Open Space to facilitate it,[3] but once you have a successful Open Space under your belt, you'll marvel at how simple it is!

You will find a really good short film about running Open Space, based around the one we ran at the Inaugural Meeting of the Transition Network at www.youtube.com/watch?v=Ux_LFjFeCvg, and the best book on the subject is Harrison Owen's *Open Space Technology: a user's guide*, Berrett Koehler Books (1993).

9. Build a bridge to local government

Whatever the degree of groundswell your Transition Initiative manages to generate, however many practical projects you manage to get going on the ground and however wonderful your Energy Descent Plan is, you will not progress very far unless you have cultivated a positive and productive relationship with your local authority. Whether it is planning issues, funding issues or whatever, you need them on board. You may well find, in many places now, that you are pushing against an open door.

It is advisable to start the process of drawing them in as early as possible in the process. Go and see the relevant people within the Council and introduce yourself and the project. It is essential to steer clear of any sense of 'them and us'. It would also be very useful to research the development plans that they have generated, to see what they have already done. Rather than your reinventing the wheel, very often Councils have done lots of community consultation and research, and although much of it will be based on dubious presumptions with regards to oil availability and climate change, it is worth checking out.

In short, seek to engage. You may well find people far hungrier for your ideas than you imagine! In Transition Town Totnes, our Liaison with Local Government Group works on ways in which the TTT Initiative can most successfully interface with Local Government. They do much that I have set out above, and seek to maximise the productive 'edge' between the two. Eventually the link with local government might extend, once an Energy Descent Plan has been produced, to someone running

World Café event at Schumacher College for local parish, town and district councillors, February 2007.

for election to the local council on an Energy Descent Plan ticket. If Steps 1 to 7 have been successfully pursued, they should get in by a landslide!

Transition Tip

For any big events you plan to run, draw up a list of people within the local authority (as well as local business, the community and so on) who you feel should be there, and invite them personally. Make sure they are personally welcomed and greeted upon their arrival.

10. Honour the elders

For those of us born during or after the 1960s when the cheap oil party was in full swing, it is very hard to relate the idea of life with less oil with our own personal experience. Every year of my life (the oil crises of the 1970s excepted) has been underpinned by more energy than the previous years. I have no idea of what a more localised society looked like in the UK; the closest I have is how towns were in rural Ireland when I moved there in 1996, where most of the shops were still owned by local families, the most memorable ones slightly damp-smelling with wooden floorboards, selling the most unusual combinations of things (paraffin lamps, boxes of biscuits and aprons) generally run by a couple in their late sixties. There is a great deal that we can learn from those who directly remember the transition to the age of Cheap Oil, especially the period between 1930 and 1960.

As part of the Transition Town Totnes Initiative, we have been doing oral history interviews with older people in the area. One, with Muriel Langford, now in her mid-eighties, contained a passage I found especially illuminating:

> "Upstairs I had Jeremy in his cot on my side, so I had an electric torch so that when he woke up I would switch on the torch and then immediately Eric would turn to the candle on his side which you couldn't have on the side where the baby was, and he'd light the candle to save the battery in the torch. We had a good little system going!"

This was in 1945, and batteries were so precious that they had to develop this system to minimise their use. Totnes at that time imported very little food, people lived at higher densities within the existing buildings (which were lived in more like bedsits rather than the larger homes they are nowadays). There was very little traffic. She spoke of moving into a flat on the High Street in 1945 which required the windows to be removed and large pieces of furniture to be winched in. This meant that the removals lorry was stationary in the middle of the road, where nothing could pass, for over 4 hours. Nowadays after 4 minutes you would have created a major traffic jam and you'd have some very irate drivers to deal with!

Transition Tip

When doing oral history interviews, avoid doing them with more than one person at a time. I recently went to do one with a lady who had fascinating stories to tell about being a land girl on Devon farms during the War, but she said, a few minutes into our chat, "My dear, I have nothing interesting to tell you at all, so I invited my friend to come along as well." A few minutes later he arrived, and I started talking with the two of them. The problem is that they began to reminisce; one would say "And down by the Quay there was that shop, what was it called?", the other would reply "Jameson's", to which the first would say, "Oh yes, Jameson's . . . now they had three sons, didn't they?" "Oh yes, Jason, he's in Australia now . . ." and so on. It was very hard to get any useful information at all!

As well as those kinds of anecdote, I find it fascinating to hear peoples' stories of how they lived then. Most people gardened – it was just what they did. People talk of the sense of community they had. It is fascinating to see, when talking to those who lived through the war years, the sense of thrift and 'enough' that those people I spoke to had. What would it take to rebuild that?

Oral histories are also very useful for getting a handle on the skills that people used to have, which links directly to Step 7 (p.163). In doing research for the Transition Town Totnes process, I found, for example, that until the early 1980s there were market gardens within Totnes, in what are now the car parks, which supplied the shops in town (as described in Chapter 3). Oral histories and historical research can offer fascinating insights into how people used to feed, employ and heat themselves. Clearly not all of it is relevant, and collecting reminiscences carries a danger of romanticising the past and devaluing the present, but there is much that can be learned.

"Several years ago, I read that elementary particles were 'bundles of potentiality'. I began to think of us all this way, for surely we are as indefinable, unanalysable, and bundled with potential as anything in the universe. None of us exists independent of our relationships with others."

– Margaret J. Wheatley (1999) Leadership and the New Science: discovering order in a chaotic world, Berrett Koehler

APPLYING THE TWELVE STEPS, No. 3: TOTNES

Totnes took a pretty straightforward path through the Twelve Steps. During the Awareness-raising stage however (2), there was no Steering Group (1), rather two or three people who co-ordinated the events quite informally, not even using the name 'Transition Town Totnes'.

Then came the Unleashing (4) and a failed attempt at forming a Steering Group (1). Then the groups began to form (5), some out of, and some independently of, the Open Space Days (6). The oral histories have been running alongside these (10), as have the other steps. Beyond the forming of groups (5), any sense that the Twelve Steps follow a chronological sequence begins to break down, other than their building towards an Energy Descent Plan (12).

Perhaps you might collect these stories together and publish them. I do think there is something powerful in making one of your first steps in doing this process to go to the elders of the community and ask for their input. It is something that in many cultures would be instinctive, but in ours has been sidelined. One interesting thing when you start doing an interview like this is that people always start by saying "I don't know why you want to talk to me, I'm sure I have nothing interesting to say to you . . .", and then go on to dazzle you with spellbinding anecdotes and fascinating information!

Oral histories can reveal a great deal about the gradual disappearance of local resilience. Here, for example, is one of Totnes's three urban market gardens, Victoria Nurseries, around the time of its closure in the early 1980s.

11. Let it go where it wants to go

Step 11 is really pretty straightforward, requiring very little elucidation. In essence, if you start out developing your Transition process with a clear idea of where it will go, it will inevitably go elsewhere. You need to be open to it, following the direction of the energy of those who get involved. If you try to hold onto the idea that it will be a certain way it will, after a while, begin to sap the energy that is building around the process. This is what is so exciting about the whole thing: seeing what emerges. It is worth bearing in mind all the way through that your role is to act as a catalyst for the community designing this transition and to facilitate people asking the right questions, rather than to come up with the answers.

12. Create an Energy Descent Action Plan

At the moment there is only one Energy Descent Action Plan, the one done for Kinsale in Ireland.[4] It makes no claims to being authoritative or comprehensive; it was, after all, done

as a student project before we really knew what we were doing. What it did very well was to suggest a template that other settlements can follow in designing pathways away from oil-dependency. An Energy Descent Plan needn't even be called an Energy Descent Plan. Other names have been proposed by those who feel the term 'energy descent' is somehow too depressing; alternative suggestions have included 'Community Resilience Action Plan' (that one didn't stick too long, for obvious acronymic reasons) and 'Energy Transition Pathway'.

Whatever it is called, the idea is straightforward. An EDAP sets out a vision of a powered-down, resilient, relocalised future, and then backcasts, in a series of practical steps, creating a map for getting from here to there. Every settlement's EDAP will be different, both in content and in style. However, they will all explore a wide range of areas as well as energy: energy descent is an issue which affects every aspect of our lives. You will also be exploring food, tourism, economics, education and a lot more besides.

Tools for Transition No. 11: How to run a fishbowl discussion
by Sophy Banks

Use this tool when you want to have a deep exploration of an issue. It allows an open forum while keeping a focused discussion. Use this for groups of minimum 10 – maybe 100, perhaps even more. The question is important – keep it open, non-judgemental and non-directive. Set a time limit – about 1.5 hours is usually plenty.

Set up five or six chairs in the centre of the room in a circle, facing inwards towards each other. Arrange further seating around this central circle, also facing inwards. Everyone starts sitting on the outer chairs. Start off the process with something that focuses people on the topic – an introduction, maybe a visualisation or reflective exercise.

When someone from the outer circle/s has something to say they can come and take an empty chair on the inner circle and join the discussion. You may have experts whom you want to stay in the centre throughout, or to start off the discussion. The discussion starts when there are two people in the inner circle. When all the chairs are occupied someone from the inner circle who feels complete should leave. One chair of the inner circle should be empty most of the time.

Only those in the centre can speak, and they speak only to each other. They may refer to what someone said earlier, but not actually say it to them or start a dialogue outside the circle. Allow some time for feedback from everyone present at the end whether they have spoken or not.[5]

The EDAP model can be seen as a flow diagram in Appendix 4. This is work in progress, to be shaped as more and more communities head off down the road of creating their own EDAPs. We have, however, identified the following ten steps in the process of creating an EDAP:

Step One: Establish a baseline This involves collecting some basic data on the current practices of your town, whether in terms of energy consumption, food miles or amount of food consumed. One could spend years gathering this kind of data, but you aren't trying to build up a hugely detailed picture, more a few key indicators around key elements of how the place functions. How much arable land is there,

how many cars come and go each day? You may well find your local Council has a lot of this information anyway. Your working groups may have identified some of this information.[6]

Step Two: Get the Local Community Plan Your local government's plans for the area are likely to have timescales and elements that you need to take into account in your EDAP. They will also be a useful source of information and data. You will need to decide whether you assume that the existing plan is based on unrealistic assumptions and will become irrelevant, or whether you want your plans to wrap around theirs.

Step Three: The overall vision What would your community look like in fifteen or twenty years if it were emitting drastically less CO_2,

WHAT ARE RESILIENCE INDICATORS?

Carbon footprinting and the cutting of carbon emissions are clearly a crucial part of preparing for an energy-lean future, but they are not the only way of measuring a community's progress towards becoming more resilient. In the Transition approach, we see cutting carbon as one of many 'Resilience Indicators' that are able to show the increasing degree of resilience in the settlement in question. Others might include:

- *The percentage of local trade carried out in local currency*
- *Percentage of food consumed locally that was produced within a given radius*
- *Ratio of car parking space to productive land use*
- *Degree of engagement in practical Transition work by local community*
- *Amount of traffic on local roads*
- *Number of businesses owned by local people*

(continued on facing page)

using drastically less non-renewable energy, and it was well on the way to rebuilding resilience in all critical aspects of life? This process will use information gathered in your Open Space Days, from Transition Tales and a range of other visioning days, to create an overall sense of what the town could be like. Allow yourselves to dream.

Step Four: Detailed visioning For each of the working groups on food, health, energy etc. (although this is trickier for Heart and Soul groups, for example), what would their area look like in detail within the context of the vision set out above?

Step Five: Backcast in detail The working groups then list out a timeline of the milestones, prerequisites, activities and processes that need to be in place if the vision is to be achieved. This is also the point to define the resilience indicators that will tell you if the settlement is moving in the right direction. Using the tool of back-

Photo by Andy Goldring

Might one good Resilience Indicator be the number of well seasoned woodpiles in a community, such as this one in Slovenia?

casting will also enable you to think through some very useful questions. An example of this is the model, being developed in Totnes, of the Local PassivHaus (see p.115).

This takes the Scandinavian model of the PassivHaus, a house which derives all its heating needs from good orientation, super-insulation and the occupants' body heat, and redesigns it to use 80% local materials. However, clearly having a Local PassivHaus built tomorrow would be a near-impossibility: a number of things would need to be in place first. There would need to be a local hemp industry in place, local lime production, people making clay plasters and, perhaps more importantly, a workforce trained up and familiar with these new materials and techniques. An EDAP offers a way of setting out the practicalities of the transition.

Transition Tip

You might commission local photographers or photography students to produce some photos of your community in 2030 to illustrate your EDAP.

Step Six: Transition Tales Alongside the process above, the Transition Tales group produces articles, stories, pictures and representations of the visioned community (such as those seen in 'A Vision of 2030' in Chapter 8), giving a tangible sense, through a variety of creative media, of what this powered-down world could be like. These will later be woven into the document.

Step Seven: Pull together the backcasts into an overall plan Next, the different groups' timelines are combined together to ensure their coherence. This might be done on a big wall with post-it notes to ensure that, for example, the Food group haven't planned to turn into a market garden the same car park that the Medicine group want to turn into a health centre. This process ought not to be too time-consuming; it is just to ensure that when combined, the different strands of the plan all tell a consistent story.

Step Eight: Create a first draft Merge the overall plan and the Transition Tales into one cohesive whole, with each area of the plan beginning with a short summary of the state of play in 2008, followed by the vision for 2030. This is then followed by a year-by-year programme for action, as identified in the backcasting process. Once complete, pass the document out for review and consultation.

If your mental picture of the final EDAP is community planning documents you have seen before, then think again. Your EDAP should feel more like a holiday brochure, presenting a localised, low-energy world in such an enticing way that anyone reading it will feel their life utterly bereft if they don't dedicate the rest of their lives towards its realisation.

Step Nine: Finalise the EDAP Integrate the feedback you receive into the EDAP. Realistically, this document won't ever be 'final' – it will be continually updated and augmented as conditions change and ideas emerge. Make a big splash when it comes out.

Step Ten: Celebrate! Always a good thing to do. In fact, you probably should have been doing this after every step above!

Transition Tip
Just because your local planning framework documents might set out plans for the next twenty years is no guarantee, in the context of peak oil and climate change, that they will come to fruition. If there are assumptions made in it that you feel to be utterly unrealistic (a new airport or hugely increased road transport), don't feel duty bound to design your plan around every word in these documents.

Beyond the Twelve Steps . . .
The Twelve Steps above set out a plan of action, and you may be forgiven for assuming that Step 12 is the end of the process. On the contrary, it is with the completion of Step 12 that your Transition Initiative really begins! The EDAP sets out the work you will be doing into the future, and in theory (no-one has got there yet) once you reach that stage, your Transition Initiative changes, and becomes, in effect, a relocalisation agency, whose job it is to implement the EDAP.

- Proportion of the community employed locally
- Percentage of essential goods manufactured within a given radius
- Percentage of local building materials used in new housing developments
- Percentage of energy consumed in the town that has been generated by local ESCO.
- Amount of 16-year-olds able to grow 10 different varieties of vegetable to a given degree of basic competency
- Percentage of medicines prescribed locally that have been produced within a given radius.

This is a new area the Transition Network is currently exploring. Your thoughts on what form other Resilience Indicators might take are very welcome. The core point is that we need more than carbon footprinting, that we could cut settlements' emissions by half, but they would still be equally vulnerable to peak oil.

Chapter 12

The first year of Transition Town Totnes

TRANSITION TOWN
TOTNES

First, a bit of background

Totnes is a town in Devon with a population of about 8,000 people. It has a reputation and a history of being quite an 'alternative' town (one wag wrote 'Twinned with Narnia' on one of the signs on the edge of the town), a reputation that began with the arrival of Leonard and Dorothy Elmhirst in 1926. She was a wealthy American heiress, and the two of them were searching for a place to develop an experiment in land-based regeneration combining arts, music and theatre. Their regeneration of the Dartington Estate adjacent to Totnes, the establishment of a college of the arts, a wide range of rural enterprises, and a reputation for Dartington as an international centre for

The historic Butterwalk in Totnes, whose name harks back to days of a more resilient economy. It was so called because on market days it was the coolest (in terms of temperature) place for sellers of dairy produce to position themselves.

the arts put the area on the map, and attracted lots of creative and alternative people to the area.

Totnes has been a hotbed of environmental activists for many years, being home to one of the most successful anti-GM groups in the country, as well as what was, for some years, one of the best LETS schemes in the country. It suffers, like many rural market towns in the South West, from high house prices, low wages and an ageing population (Totnes is the town in the UK with the highest per capita number of people over 60 living on their own). It has a vibrant market on Fridays and Saturdays, and a strong culture of local food. It is unusual in that it has largely managed to avoid the Clone Town Britain phenomenon, with an uncharacteristically high proportion of small, locally owned shops. I moved there from Ireland in September 2005. I could be telling you the stories of any of the other developing Transition Initiatives here – Lewes, Penwith, Bristol (indeed I will go on to tell those stories in less detail) – but Totnes is, in many ways, the flagship initiative, and also the one I am most familiar with.

Then, a bit of prehistory

What follows is an attempt to write a history for a project which is very broad, diverse and dynamic, and is still ongoing and evolving. In the interests of conciseness, some things have been deliberately left out, and unavoidably, given how big it has become so fast, some things I don't even know about may have fallen through, but hopefully, this will give you an overview of this one town's story.

Transition Town Totnes (TTT) began in October 2005 with a screening of the film *The End of Suburbia*. Over the next few months, fellow peak oil educator Naresh Giangrande and I held a series of talks and film screenings and began to network with existing groups. At this stage the process wasn't called 'Transition Town Totnes'; indeed it didn't come under any collective name until the week before the Official Unleashing. We tried to make any events that we ran as much about building networks and

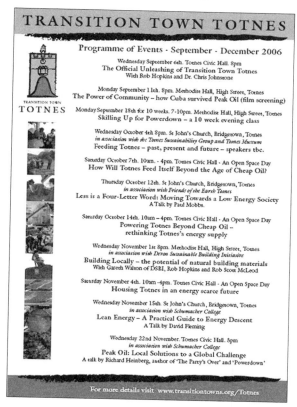

The very first Transition Town Totnes programme of events

relationships as about viewing films. You can see some of the tools and exercises we utilised on pages 154-5.

September 2006

Early in the first week of September, posters began popping up all over Totnes advertising an 'Unleashing', no doubt leading to many people wondering what on earth such a thing might be. All was revealed on Wednesday 6th September at the Civic Hall in Totnes, when the 'Official Unleashing of Transition Town Totnes' took place. The evening was attended by over 350 people, who packed the hall to hear presentations by myself and Dr Chris Johnstone. It was opened by the then Mayor of Totnes, Pruw Boswell, who talked enthusiastically about the initiative and her hopes for its success. "If this can happen anywhere," she told the audience, "it will be in Totnes."

I spoke about peak oil, what it is and why it is such an urgent issue. I talked about the concept of energy descent and of relocalisation, and told the story of the Kinsale Energy Descent Action Plan. I set out my vision for how the Totnes process might unfold, and what TTT's plans were for its first four months.

Chris Johnstone talked about how insights from the addictions field can inform energy descent work. He argued that peak oil and climate change represent our call to power, our defining moment, and by engaging with them, we will find a great deal of inner resources and strengths we didn't know we had. At the end of his talk he divided the audience into pairs and got them talking to each other.

In pairs they focused for three minutes on each of the following questions: "When I hear about climate change and peak oil the concerns I have include . . .", "My vision for Totnes after peak oil is . . .", and "The steps I can take

"A good start – looking forward to getting involved in a practical way."

"A much clearer idea of the whole situation. Feeling inspired and excited. No longer feeling so impotent – understanding raising awareness is as important as knowing what to do."

"Very inspiring. I look forward to the unfolding."

"Excellent. Good to hear the depressing news but great to be inspired, empowered and excited."

"Brilliant! I'm really excited about this, count me in!"

"Inspiring, informative, exciting, hopeful, action focused and well structured. Thank you."

"Excellent evening. A great start to what will be a ground-breaking project. Thanks for your vision!"

– some feedback from the TTT Unleashing

towards this vision are . . ." After this, people were invited to write down their thoughts on these questions, each on a different-coloured post-it note, which were then stuck up on the wall so people could read them.

People remarked afterwards that they really enjoyed the opportunity to discuss their thoughts, and to chew over what they had heard. People left the evening very enthusiastic about the process. It was designed from the outset to be an historic occasion, the night that people would look back on as the start of the whole process, and it very much had that feel. The Unleashing generated a lot of energy that has really propelled TTT forward since.

A few days later we showed, for the second time, the film *The Power of Community* which relates the experience that Cuba underwent when it experienced its own oil peak, albeit an externally imposed one, when the Soviet Union collapsed in 1990. Despite having sold out the previous venue the first time we showed the film, we once again filled it, with over 100 people and a very vigorous discussion after the film.

Around this time, the evening class, Skilling Up for Powerdown (see page 194) began for the first time. The group of 35 people from a range of backgrounds began the course which covered a wide range of subjects, and was designed to train up fieldworkers for the project.

October 2006

October began with an evening event called 'Feeding Totnes, past, present and future' held at St John's Church, which featured, appropriately, three speakers. The first, Mary Bartlett, gave a fascinating account of her years as a horticulture student at Dartington in the 1960s, and the second, Helena Norberg-Hodge from the International Society for Ecology and Culture, gave an overview of the dangers of globalisation, and how communities are strengthening their food economies around the world in response. The last speaker was Guy Watson from Riverford Farm, one of the UK's biggest organic farm businesses, who talked about how the farm (situated close to Totnes) runs, and the steps they are taking to make it less oil-dependent.

This was followed three days later by the first TTT Open Space, on food. This was attended by well over 100 people, although people came and went during the day (the practicalities of Open Space are explored on pages 168-9). This was the first time we tried to document an Open Space in real time on the Wiki website. It went very well, and proved to be a very potent tool. The day generated a great deal of networking and subsequent projects, such as the Seed Swap day, emerged from discussions on the day.

On the 12th, Paul Mobbs, author of *Energy Beyond Oil*, gave a well-attended talk which set

> *"The whole point about modern society in Britain is that the only way we know to have fun these days is to expend energy. We need to re-learn all those skills our grandparents had for having fun with very little. That's what it's really going to be about. If you look at human history, we can bear anything, so long as we have a bit of fun at the same time. The fun bit is what we really have to work on."*
>
> – Paul Mobbs, author of *Energy Beyond Oil*, interviewed in Totnes, October 2006

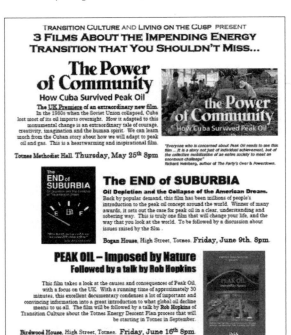

out a rich overview of the energy challenge facing the UK, which was again followed by an Open Space Day on energy. Although a few less people came than had attended the Food day, it was still a vibrant and productive Open Space.[1] On the 26th we had a talk, arranged at the last minute, by organic gardening expert Bob Flowerdew. Bob drew a large crowd – quite a different demographic than previous talks – and passed on many insights from his knowledge about gardening.

On October 17th, the Heart and Soul group had their launch. The Heart and Soul group set out to explore the psychology of change, and identified the questions it aims to address as follows:

> "What is it actually like to be human, alive and awake at this time in the history of the planet, the history of humanity? What is it really like to live within, and inevitably be part of, our current global economic system? What of our dreams and visions, fears and angers, grief, passion and inspiration? What motivates us to creatively transform how we live, and equally, what tends to 'shut us down' and reach for comfort instead? How are we shaped by the society in which we live, and how does what lies within us, in turn, create the society we make together? Perhaps most importantly, what structures or processes do we need to create to support us in both the practical work ahead, and in the 'change of heart' on which that work rests?"

Their launch evening was very well attended, with singing, reflection and discussion, and they have gone on to be one of the best attended of the groups. As their programme of events emerged, other more experiential things were added, exploring a variety of different tools.

Another group that kicked off around this time was the Energy group, which had its first

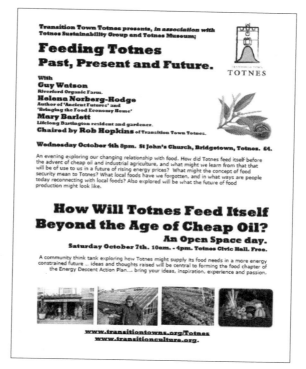

meeting in the shed at the end of a garden, where, Tardis-like, an extraordinary number of people managed to fit in a very small space and explore the energy aspects of TTT. The other group that emerged in October was the Health and Medicine group, whose first meeting was a small but dynamic affair, exploring what such a group might go on to do.

November 2006

November's first event was an evening about building, entitled 'Building Locally; the potential of natural building materials'. Three speakers discussed different aspects of building and construction: Gareth Walton introduced the Devon Sustainable Building Initiative, Jim Carfrae discussed the process of building the straw-bale house he created in Totnes, and I spoke about the range of natural building materials and techniques. This was followed three

PETER RUSSELL ON TRANSITION AND CONSCIOUSNESS

Do you think there's a danger in emphasising consciousness change above practical action, that some people may think that actually we can just evolve magically out of this crisis, that the universe will somehow sort it out for us, we don't actually need to get our hands dirty ourselves?

"I see there's a big danger in making consciousness change the top thing or the priority. It's going to take a lot of physical work, social changes, economic changes, many sorts of changes which are going to require work, adjustments and decisions, in the active world. Those things are going to happen more fluidly, more easily, and I think more constructively, if at the same time we are freeing our consciousness up from the old mode, that old, self-centred, materialistic mode, which actually created the problems in the first place on a larger level.

If we don't also tend to our consciousness, our mind, our psyches, then we're going to be repeating some of the same mistakes. I think the opposite error is equally important to avoid — which is purely looking at what must be done, without bringing in the consciousness element. That's the way we normally

(continued on facing page)

days later by our last Open Space of 2006, entitled 'Housing Totnes in an Energy Scarce Future' which was, again, well attended, and generated a wealth of ideas.

On November 8th, Simon Snowden of Liverpool University, who has been developing the tool of Oil Vulnerability Auditing (OVA) made his first visit to Totnes. Simon ran a one-day workshop for ten local businesses at the South Hams District Council offices, outlining the OVA process and the insights it could offer to businesses in terms of understanding their degree of vulnerability, and therefore, risk.

Simon Snowden of Liverpool University introducing the concept of Oil Vulnerability Auditing to an audience of local businesspeople.

On November 9th, the TTT Arts Group had its launch. A group of local artists came together and discussed how their art practices could inform, document and inspire the TTT process. Although trying to get a lot of artists to do things together has been likened to 'herding cats', there was an energy generated to drive forward this aspect of the project.

Around this time, Schumacher College hosted a remarkable course called 'Life After Oil', a two-week intensive immersion in peak oil. One of the teachers was David Fleming, economist and originator of Tradeable Energy Quotas, who on November 7th gave an evening talk for TTT called 'A Practical Guide to Energy Descent'. To a packed hall, he wove a charming tapestry of subjects, from peak oil to choral music and from carbon rationing to Wordsworth. The Transition, he argued, is as much about reclaiming and rebuilding culture as it is about solar panels. He ran wildly over time, and everyone loved it. The previous evening he and I had sat, together with Stephan Harding, on a panel to discuss the film *An Inconvenient Truth* at the Barn Cinema,

Dartington. Someone asked the panel, "What gives you hope?" David replied, after some deliberation, "Bach".

Later that week I visited the offices of a local film production company. During our discussions about localisation and so on, they produced, from the wall in their office, a framed 1810 bank note (the building used to be a bank). It got my mind ticking about local money and bank notes. Also around this time, a group began to meet to discuss the creation of a local food directory. Inspired by one produced in the Forest of Dean, they began to explore how such a Directory might be created for the town.

Another group that began to emerge in November was the Local Government Liaison Group, made up of local councillors and others who wanted to connect TTT with local government. In the TTT email Bulletin, they wrote:

"The proposed TTT Local Government Liaison sub-group would aim to promote the development of local government policies that would encourage and avoid hindering the action necessary to reduce the fossil fuel dependency of Totnes and its environs in line with the proposals arising from other sub-groups and the TTT process."

They rapidly became a key aspect of the work of TTT. The final event of November, on the 22nd, was a talk by Richard Heinberg at the Civic Hall, attended by almost 400 people. People travelled from as far away as Leeds to hear Richard's talk, which was very well received. It was, alongside the Unleashing, our largest event of 2006.

December 2006

A lot was starting to happen around this time with the Totnes Renewable Energy Society (TRESOC). This is a sister project to TTT, which emerged from the Energy Open Space Day and aims to allow local people to invest in renewable energy infrastructure. Also emerging at this time was the Economics and Livelihoods Group.

December 15th was the first time the Project Support Group met. A previous attempt at forming a Steering Group had not really worked, but using this model it was possible to create a group to co-ordinate and guide the process whereby each of the working groups contributed one person, making it more dynamic and vital. This settled in very well and has been a very effective decision-making tool.

January 2007

The first event that we ran in January was a training day with Andy Langford and Liora Adler of Gaia University, on the theme of Designing Productive Meetings. It

had become clear as people came forward to facilitate the various TTT groups that, although they all have a great deal of enthusiasm and passion for the work, they may not necessarily have the skills in terms of facilitating meetings, conflict resolution and so on. This workshop brought together these various group facilitators for a very useful day exploring these skills.

On the 17th, we hosted a talk by Peter Russell called 'Waking Up in Time'. It was the biggest crowd we had yet drawn to St John's Church, nearly 300 people. Peter talked about the importance of consciousness in the Transition, something we hadn't touched on yet.

Sunday 28th was 'Seedy Sunday', the TTT seed exchange, with hundreds of people bringing out seeds – the weird, the wonderful and the heirloom – to swap with other people. The day also included talks and demonstrations, and was a very energetic event. TTT also, at this time, got its first mention in the national press. The Guardian's Eco-Sounding section ran a brief piece as follows:

Rummaging through seed packets at TTT's Seedy Sunday.

Photo © Frankie Wellwood

approach things – we see a problem out there, we must fix it, change it, do something about it. We have to bring in the mind – what is it in our thinking that's causing our problems or stopping us solving them in the right way?

So there definitely needs to be a balance, the two supporting each other, but a shift in consciousness is going to support the work we have to do on the outer. I don't believe that if we all just meditate and sit back, the world will miraculously sort itself out. I don't believe that at all. It's definitely going to need a lot of very challenging work, challenging decisions, which is going to probably push through some personal discomforts, hardships maybe, as we make adjustments. That's not going to be easy, and that's why I think we're going to need to be looking at our consciousness at the same time so that we can ease our way through that process."

– Peter Russell, interviewed in Totnes, 18th January 2007

*TONY JUNIPER ON PEAK OIL
AND ENERGY DESCENT*

"I think there will be a transition, and I think it is pretty impossible for us to have an orderly withdrawal from the Carbon Age that happens very quickly, we can't do it. Our infrastructure, our transportation systems, our fuel mix, our agriculture crucially, everything, is geared up to be heavily dependent on fossil fuels. It will take a while to get out of it, but the quicker we start it the sooner we'll do it, but also the more orderly the transition will be.

This mixing up of decarbonisation with a shock built around the rapidly rising price of oil will be harder to cope with. If we start now and begin to decarbonise, with all the technological things we already have, from the bicycle to concentrating solar power, all that stuff already exists, we need to get it moving and get it into the market fast, so we can start the process while we still have the economic stability and the money and the social comfort to do this without even noticing it.

(continued on facing page)

"The idea of Totnes, Devon, running out of oil conjures images of matrons hitch-hiking to Exeter and retired chaps cooking on open fires, but the town is deadly serious about producing local food as and when the black stuff runs out. It has now declared itself Britain's first 'transition' town, and permaculture guru Rob Hopkins is drawing up a 25-year plan to see how Totnes could support itself without oil. There have been meetings about how food and energy could be produced locally, but the plan is to set up a local energy company, rewrite the local development plan and persuade others to join the movement. It must be working. Stroud in Gloucestershire and Lewes, East Sussex, have just joined."

February 2007

The Transition process in Totnes continued to deepen, with the Project Support group meetings going well, continuing to explore structure and so on. During February, Schumacher College ran a three-week course on climate change, with some excellent speakers. We borrowed some of them for some public events that we co-presented with South Hams Friends of the Earth, Schumacher College and TRESOC.

The first of these was Aubrey Meyer, the creator of 'Contraction and Convergence', and the second was by Tony Juniper, the director of Friends of the Earth. Both were very well attended. While Tony was at Schumacher College, he also participated in an event that had been organised by the Local Government Liaison group, a World Café day for local councillors. World Café is similar to Open Space, but it can be more guided, less chaotic, and self-organising (you can read how to run a World Café event on pages 184-6). The event was attended by 23 councillors, from parish, town and district levels, as well as by the local MP, Anthony Steen. The event began by explaining peak oil and climate change, and then suggesting that, at present, local government decision-making is based on the assumptions that oil prices will remain low for the foreseeable future, that climate change won't impact for some time, and that the move away from people doing things for themselves will continue indefinitely.

We asked what their forward planning might look like if those weren't the assumptions. A wealth of thoughts were generated, on subjects as diverse as tourism, planning, transport and the economy. It was done using Chatham House Rules, meaning that what is said in the meeting can be cited outside, but is not attributable to any individual.

A few days later, the first Totnes Pound (of which more later) went to the printers, designed with one side being a facsimile of the 1810 Totnes

The original 1810 Totnes banknote spotted on the wall of a local film production company, which went on to inspire the first modern Totnes pound note.

banknote I had seen hanging in the old bank three months previously.

Two other notable events in February were the Heart and Soul Open Space Day, exploring all the different elements of this aspect of TTT, which was very well attended and did a great deal to shape and inform the work of that group, and also the first meeting of the Transition Tales group. You can read more about the work of this aspect of TTT on pages 200-1.

March 2007

Early March was a Totnes Pound frenzy, trying to get the notes printed and ready for their launch on March 7th. In the short amount of time available, eighteen shops were signed up for the scheme, and the notes were designed and printed. The first note was given to Totnes Museum, and the next ones to the members of Totnes Town Council.

The Council had invited me to speak to them about TTT, to bring them up to speed with progress. At the end of my talk I told them that although I didn't make a habit of giving money in brown envelopes to Councillors, for them I would make an exception (the notes had come from the printers wrapped in brown paper), and proceeded to give them each one of the first Totnes Pounds. Following the meeting, the Council voted unanimously to endorse the work of TTT, which was a great boost to the project.

The notes themselves were launched at an event called 'Local Money, Local Skills, Local Power' held on March 7th, which also included a talk by Molly Scott Cato of gaianeconomics.org, an economist and member of the nascent Transition Stroud project. The Pound got some great publicity, our local MP Anthony Steen being photographed brandishing fistfuls of notes. This was then followed by an Open Space Day on economics, which generated lots

Local residents proudly displaying the brand new Totnes Pound on the night of its launch.

of ideas about the role of the Pound, alongside other tools such as LETS, Time Banks and Credit Unions.

Also in March we organised a tree-planting on Vire Island in the centre of Totnes, to plant five almond trees and two walnuts. The then-Mayor, Pruw Boswell came down and planted one of them, and we got some great photos in the paper to launch our 'Totnes, the Nut Tree Capital of Britain' initiative.

The next talk in our series of events was by Jerry Mander, one of the co-founders of the International Forum on Globalisation. His talk

Photo © Arthur Kay

Former Totnes Mayor Pruw Boswell balances walnuts on a shovel at the first TTT nut tree planting event.

The place we could finish up could be so much nicer than the one we've got now! That's the other thing that's crucial to get across to people, we're not heading back to a new Stone Age or a Dark Age, we're heading towards a much brighter, more secure future, where communities are rebuilt, pollution is a thing of the past, we've got food security, biodiversity, people have long comfortable lives, energy is secure for ever. Getting that picture across is very difficult to do, because the only thing you can ever paint an accurate picture of is the past of course.

When people look back to the pre-oil age, the pre-carbon age, it was miserable. People died at young ages, there were diseases, there were not very high levels of social comfort, and the tendency is to equate that with the future, whereas in fact the future could be very different indeed. We do have to paint that positive vision, but it's not easy because we haven't been there, so you can't show people a film of it, but they're the kind of pictures we have to create."

— Tony Juniper, interviewed in Totnes, February 2007

Tools for Transition No. 12: Running a World Café Event

World Café has been summarised as being about "awakening and engaging collective intelligence through conversations about questions that matter".[2] It is a very powerful tool for exploring specific questions. It differs from Open Space in that it is less chaotic, and can offer a powerful way of exploring specific questions and issues. It is based on the idea that for many people, the place where the richest conversations take place are places where they feel relaxed: at a table – be it a kitchen table or a table in a café – with a cup (or glass) of something in their hands and perhaps some nice biscuits.

It is a tool we have used to bring people together to ask very specific questions, rather than, as in Open Space, allowing them to set the questions. World Café also differs from Open Space in that it requires more organisation in advance, but if done successfully, it creates a very potent and memorable space for exploring the questions you have devised.

The World Café, a global network of practitioners, have devised seven principles of World Café, which set out the whole process very clearly:

1. Set the context

One of the keys to a successful World Café is that you have prepared the event well. This involves giving consideration to what the topic for the Café will be, how you will frame the question(s) that will be explored, who should be there and how you will invite them, where and when it will be, and what outcomes you are hoping for from the event.

2. Create a hospitable space

You need to make the place where you are doing your World Café event as hospitable as possible. Some places lend themselves more naturally to a relaxed ambience than others. A bright sparkly conference centre or a motorway service station probably don't really do it, while somewhere that has a more relaxed intimate feel would be much better. It needs to feel like a safe space, somewhere people feel comfortable both physically and in discussing ideas.

Creating this space starts early on with the invitation you send to people. The invite should be bright and bold and stand out from the rest of the post people get, and should include the question(s) that the event will explore. When arranging the room, set out round tables, of a size that will seat about five people, in a random arrangement, and then use large paper tablecloths. Each table should also have a supply of pens, some flowers and a candle.

One of the other keys to the success of the whole thing is the availability of food and drink. There should be some of both available when people arrive, and then also available throughout the session, so people can just wander up and help themselves. Some wine and juices, and a mixture of sweet and savoury things goes very well.

3. Explore questions that matter

As I mentioned above, the question that you ask needs a lot of consideration. The framing of your question will make or break your event. You can either have one overarching question which is explored in increasing depth in the progressing rounds of the event, or a number of questions that explore different aspects of an issue. Your question should be relevant to the group, it should be clear, thought-provoking and invite deeper reflection. It should invite the exploration of possibilities and connect those present to why they came.

A World Café event I was part of, which was facilitated by Second Nature[3] at the Cultivate Centre in Dublin, was designed with one particular objective, "to plan how communities can thrive and prosper after peak oil and through energy descent". To do this, we devised four questions, each of which was the subject of a round of conversation. They were:

Participants in a World Café session at the Cultivate Centre in Dublin using Mind Mapping to record their ideas.

- "Is peak oil a crisis or an opportunity for communities?"

- "How relevant is the energy descent model to me and my situation?"

- "How do we communicate community responses to peak oil, both within and between different communities?"

- "What are the skills we need to learn and the training and education we need to put in place to respond to peak oil?"

4. Encourage everyone's contribution

Everyone who attends a World Café event becomes part of a larger system, a living network. The idea of World Café is that it aims to maximise the number of connections made between people, understanding that the more connections there are, the more collective intelligence is unlocked. This is done by every fifteen minutes, a bell (or similar) sounding, which means it is time to move to another table. Over the space of a few hours, it means that you actually get to meet most if not all of the people in the room, and exchange ideas and thoughts with them.

5. Connect diverse perspectives

This is achieved by ensuring the maximum of mixing and cross-pollination, which creates a dense web of connections. Each time the bell sounds and people move to another table, they bring with them the threads of conversation from where they were to a new group of people. At the beginning of the process, each table

Tools for Transition No. 12 (continued)

selects a Table Host, who remains at that table throughout. It is the Table Host's responsibility to scribe the points raised in each conversation on the tablecloth, so as to create an accurate (and legible) record of what was discussed.

Each time the groups change, the new session begins with the Table Host sharing what has already been discussed at that table, and the new people briefly share what happened at the tables they were on previously. This ensures the maximum weaving together of this diversity of threads.

6. Listen together and notice patterns

Key to a successful World Café is the art of good listening. It is an art that many of us are not that experienced at, but it is a learnable skill. Listening is more than just being quiet with your ears open. According to theworldcafe.com[4] there are certain features to good listening:

- Listen to what the speaker is saying with the implicit assumption that they have something wise and important to say

- Listen with a willingness to be influenced

- Listen for where this person is coming from and appreciate that their perspective, regardless of how divergent from your own, is equally valid and represents a part of the larger picture which none of us can see by ourselves.

- While you are speaking, it is good also to try to be clear and succinct, and not hog the space. As a listener, it is good to try to avoid planning your response to what the speaker is saying, and to listen with openness. Try to listen for what is being spoken 'under' the words, what deeper patterns are emerging and what new questions are emerging.

7. Share collective discoveries

At the end of the process, the results of the event can be shared in a variety of ways. Firstly, all the tablecloths can be pinned up for all to see. Then you might have a 'go-round', where each Table Host summarises the main conversation points on his or her table. This could then be followed by a more general 'go-round' to give people an opportunity to share reflections on the process, how it went for them, and what deeper questions were raised. This process can also be continued by typing up the sheets and emailing them out to everyone a few days later, as 'minutes' of the discussion.

So there you have it, ambience, good food, conversation, lots of mingling and an outpouring of ideas; World Café in a nutshell![5]

was called 'At the end of the era of globalisation: turning again to the local'. He gave a very thorough overview of globalisation and why its lifespan is limited, and the potential of a more localised future.

April 2007

On April 5th, TTT presented, at the Barn Cinema in Dartington, the South-West première of the film *A Crude Awakening*. The evening was a sell-out, and the film, an excellent motion picture which does for peak oil what *An Inconvenient Truth* does for climate change, was very well received. It was followed by a discussion with a panel of local people with an interest in the subject.

On April 12th, one of the most dynamic groups to emerge from the TTT Arts group, the Sustainable Makers Project, had its first meeting. It brought together a collection of craftspeople from a range of disciplines, to look at how best to support the concept of useful crafts made locally.

On the 19th, I had a very odd call from a reporter in Mexico, who had somehow heard about TTT and wanted to write a piece about it. However, her main question, to which she kept returning, was: "So, to what extent is Transition Towns like John Lennon's 'Imagine'?" I mean, how would you answer that? It led to a bemusing piece in the Mexican paper, and a rather amusing piece in the *Totnes Times* (see below).

On April 22nd, TTT ran its first Transition Tales event. The idea was to bring together storytellers and writers, as well as anyone with creative urges, to begin to tell stories around different points in the town's transition. In order to do this, a few days in advance of the event a few of us created a 'timeline', a chronology of the next 25 years, with key events such as oil peak in 2010, carbon rationing being introduced in 2012 and the first car park being returned to being a market garden in 2015, among other things. People were then invited to write newspaper articles from any point along the timeline. The event generated some hilarious and also some rather touching stories. We giggled a lot together whilst, I suspect, also exploring some quite intense ideas and possibilities.

Confusion in Mexico

TRANSITION Town Totnes is big news down Mexican way – but the question is why, and how do they know about it?

The Mexican newspaper Excelsior has a three-quarter page article on the organisation which is preparing the town for life after oil.

And Mr Transition – Rob Hopkins – and former town mayor Cllr Pruw Boswell e-mailed quotes to be used in the story.

'I know they have used it because I can see my name there,' said Cllr Boswell. 'But I can't understand what it says because it is in Spanish. And Rob can see his name there but he can't understand it either.'

She added: 'The thing is how did Mexico find out about Transition Town Totnes?'

The mayor is also baffled by a map of England at the top of the page which shows Liverpool, London and Totnes as if they were the only places that mattered.

From the Totnes Times, *May 2007. Reprinted with permission.*

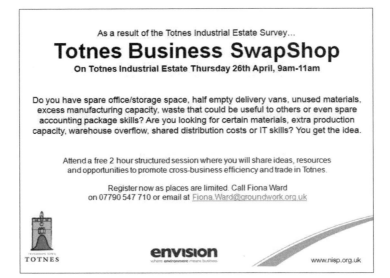

TOWN POISED FOR ITS OWN CURRENCY

"A south Devon town has taken a step towards having its own currency after a month-long experiment.

Three hundred Totnes Pounds were printed in March for circulation only in local outlets. Eighteen shops joined Transition Town Totnes (TTT), a new group campaigning for a more self-sufficient community. The experiment, which has just finished, could be followed up by another print-run of 3,000 notes later this year.

TTT based the idea on a similar scheme in the Southern Berkshire region of Massachusetts, US. There the alternative currency, Berkshares, can be swapped for dollars in banks. In Totnes TTT gave away 300 notes at a public meeting in March. Marjana Kos, of TTT, said: "It's keeping wealth here. It's keeping local trade alive and supporting local businesses."

(continued on facing page)

A couple of days later, as part of the Transition Tales Initiative, we hosted an afternoon event at Schumacher College for teachers, to ask for their thoughts on the Transition Tales project so far, and to explore with them how it might fit in with their work in schools. Both primary and secondary schools were represented, and the discussions were very valuable in shaping the subsequent work.

Around the end of April, TTT presented Totnes Library with over £1,500 worth of books, which had been individually sponsored by local shops and individuals, as well as some DVDs, which can be rented for free from the local DVD shop. The end of April also saw, after some delay, the beginning of the TTT Building and Housing group. Their first meeting, which was very well attended, was basically a 'getting to know each other' meeting, and also got the group starting to think about what it might like to do.

May 2007

On May 1st, TTT ran a 'Business Swap Shop' on Totnes Industrial Estate, which was the first such workshop, getting groups of businesses together to look at what they need and what they throw out, based on the principle that one person's waste is another's raw material. A diversity of businesses came together, and found that there were indeed a lot of connections that could be made. That week, thanks to the kindness of a local resident who offered to fund some aspects of TTT, we were able to advertise our first paid post.

Keith Ellis from Moving Sounds[6] and Transition Town Lewes[7] came to Totnes to run a one-day course on making short films using a digital camera and a laptop and posting them on YouTube. One of our only disappointments with the TTT process up to this point was that no-one had been documenting it on film. Several production companies had wanted to, but had, for one reason or another, been unable to attract funding, or wanted to take an angle on it that we were unhappy with, so all we had were photos and notes. The beauty with Keith's approach is that it makes it very possible to create short films to document different aspects of TTT's work. The course produced four short films and was great fun. You can see some of what they produced by going to YouTube and searching for Transition Town Totnes.

On May 8th, international permaculture teacher Penny Livingston-Stark came to Totnes, as part of a course at Schumacher College, and we put her on to speak as part of our programme, an evening called 'Permaculture: designing for transition'. Her ideas about the power of permaculture as a design tool for post-carbon societies were very insightful.

On the 15th, an evening meeting was held in

Totnes called 'Funny Money: where next for the Totnes Pound?', which explored the possibilities of a second phase. It drew a lot of people together to explore the possibilities of an enlarged scheme, and also brought out a number of new volunteers for the project. The first scheme still had a couple of weeks to run, but they wanted to look beyond it to what might follow it.

The next big evening talk was on the 23rd, and explored 'New Strategies for Zero Energy Housing'. The speakers were Bob Tomlinson of the Living Villages Trust and Bill Dunster, the designer of BedZed in London. The evening was very well attended and generated a lot of discussion around what might be appropriate strategies for Totnes.

On the 26th we held the 'Totnes, Nut Tree Capital of Britain, Design Day'. Sixteen people came together to explore where in Totnes might be most appropriate for nut-plantings. A couple of speakers opened the day: Martin Crawford from the Agroforestry Research Trust[8] was one of them, and he explained that walnuts were not a good choice for urban tree plantings, unless you wanted a major ongoing squirrel cull! He suggested sweet chestnut and heartnuts as better choices. The group split into two, and armed with maps, 'surveyed' the centre of Totnes looking for good spots. May 31st was the last day of the first Totnes Pound scheme. The shops were interviewed, the last notes gathered in, and an avalanche of media coverage was dealt with. BBC Devon news ran a great piece on it, and the coverage reached the pages of the *Daily Mirror*, and even the *Buenos Aires Herald* (maintaining our levels of South American coverage!) You can read more about the Pound below.

June 2007

The next event on the programme was a talk by Andrew Simms and Nic Marks of the New Economics Foundation, called 'The New Economics, from growth to wellbeing'. Andrew's talk focused on the challenge of climate change, and Nic looked at the concept of Happiness Indicators, new ways of looking at progress, and how despite increasing national wealth, our sense of how happy we are has fallen steadily since the 1970s.

On the 11th, two of our biggest events thus far took place in the Great Hall at Dartington. The first ran during the day and was called 'Estates in Transition'. The idea was that for a town like Totnes to explore the practicalities of Transition is fairly futile without the re-engagement of the land surrounding the town.

Local landowners and estate managers discussing the implications of energy descent for their management decisions in a World Café setting.

Louise King, manager of the Riverford farm shop in Totnes, said: "We like our own products, so it just seems right to have our own currency." What might seem like criminal activity is actually perfectly legal – the notes were copies of a pound note last in circulation in 1810. Paul Hall, of Colour Works printers, who printed the notes, said: "We thought it was a little bit strange. Printers aren't usually asked to produce currency, but once it was explained that it was a reproduction of a Totnes Pound as opposed to a Bank of England pound, we were happy to do that."

The pound note, which has a facsimile of the old pound note on one side, has a list of participating outlets on the reverse.

The note also shows how many times the note had been exchanged with a tick box. TTT says the only objections have come from people who thought it was too big and it will be producing a smaller, more difficult to forge note if members agree to a further print run later this year."

– BBC News Online, June 4th, 2007

Historically there would have been a lot more links than there are now. Estate managers, trustees and CEOs were invited from across the South-west, and the day featured a mixture of presentations by people such as David Furzedon, head of the Country Landowners Association, Chris Skrebowski, editor of *Petroleum Review*, solar expert Jeremy Leggett, Patrick Holden of the Soil Association and myself, and Open Space-style group discussions.

The day was a very powerful exploration of the issues, and introduced the work of TTT and the concepts underpinning it to a wide range of people. That evening, also in the Great Hall, was an evening called 'Food and Farming in Transition', which made some of the key speakers from the day available to the public. The speakers were Chris Skrebowski, Jeremy Leggett, Patrick Holden and Vandana Shiva. The evening sold out the Great Hall, with nearly 300 people, and more turned away. It was an energising exploration of the impacts peak oil and climate change could have on agriculture, and a look at what agriculture may go on to look like. During this time, plans were also afoot for the second release of Totnes Pounds. Drawing from the observations from the first one, the second one would be more available, smaller and more plentiful. The

design work was completed, and they went to press at the local printer, incorporating a number of clever security features on each note. They were launched on the 20th (in fact they were launched three times).

The first launch of the Totnes Pound was at the relaunch of the Totnes Chamber of Commerce (one of the sponsors of the notes), in which TTT featured heavily. Then, later that evening, the Pound was also launched at a talk at Schumacher College by economist Wolfgang Sachs. Particularly at the Schumacher talk, there was a huge amount of enthusiasm for the project, and when I went home afterwards my head was buzzing with the potential of the whole thing, to the point where I actually found it hard to get to sleep.

Two days later, the Pound had its final launch alongside the launch of the Totnes Local Food Directory, which had emerged from the hard work of the food directory group, published under the heading 'A Celebration of Local Food'. It combined the Directory itself, articles, recipes and work by local artists. The launch was possibly the wettest day of the year up to that point (June 2007 was the wettest on record), in a rain-swept Civic Square. The Town Crier opened the proceedings, and then the new Mayor of Totnes spoke about the need for more local food and officially launched the Directory and the Pound. There were also talks about the Pound and the Directory, followed by a specially composed song about local food by a local choir group. The group then spent the rest of the day giving out food directories on the High Street.

That week also featured the launch by the Heart and Soul group of the Home Groups, an initiative which could potentially play a big role in the future evolution of TTT. The Home Groups concept is explored in more detail below, but the evening was very positively received, with two new groups forming, and one that was already up and running sharing its experience of how helpful they had found the process. July finished with a team from Radio Scotland coming down to interview many of the key people in TTT for a long piece they wanted to do on the Transition idea.

The last important development to take place in late June was the Oil Vulnerability Auditing training run by Simon Snowden in Totnes. As already mentioned, we had been planning this training for some time, and sixteen people undertook Simon's two-day training. The training went very well, and some of those who undertook it will go on to do three pilot audits in July. TTT is the first group to run training in this approach.

July 2007

July opened with our local MP, Anthony Steen, writing his monthly column in the local press about TTT, peak oil and relocalisation. In it, he mused:

> "When cheap oil is no longer, what will happen to our lifestyle? With climate change leading the global action agenda, we must now consider 'Life after Oil'. Could less oil result in a renaissance of British agriculture, and expansion of small businesses? Could peak oil help change attitudes and rebuild 'social capital' by bringing people and disparate groups together, united in tackling a common world problem?"

Speakers for TTT during July included Alastair McIntosh, author of *Soul and Soil*,[9] who discussed land reform in the context of Celtic spirituality, and Marianne Williamson, who added a spiritual dimension. As part of TTT's post-John Croft decision to weave more celebration into its daily fabric, we held the TTT Summer Picnic

Photo © Nevenka Mulej

Noni McKenzie and new Totnes Mayor David Horsburgh launching the Totnes Food Directory.

by the River Dart on a very pleasant summer evening (the same evening in fact that most of the Midlands were disappearing under water in the heaviest rain on record). Food, conversation and rounders went on late into the night.

On July 19th, at Bowden House near Totnes, the TTT International Youth Music Festival, which brought together students (mostly from Spain) who were studying at the Totnes School of English and local kids for an evening of music, circus skills and other entertainment. The theme of the evening was TTT, and they had painted a glorious day-glo banner of the TTT logo. The evening was solar-powered, and was as much a cultural exchange as a party.

The last day of July was very special. After weeks of legal t-crossing and i-dotting, TTT finally had its own office for the first time. The July 31st meeting was the first we had there; it felt good to have put down roots in the centre of the town.

August 2007

August is the month where we try to take a

Local and international exchange students performing a play about the history of oil at the TTT International Youth Music Festival.

break, and in the main, apart from organising the next programme of events, we managed to achieve this. One of the few things that happened was the broadcast on BBC Radio Scotland of the item recorded a month or so previously in Totnes. The resultant piece was one of the best pieces of media anyone had yet put together about TTT.[10] It was followed by a phone-in, where pretty much everyone who rang in said they thought it was a wonderful idea, or that they were already underway in their communities starting a Transition Initiative.

September 2007

It was heartening to discover early in September that, according to the Style Magazine in the *Sunday Times*, "community spirit is back in fashion". An article called 'Love Thy Neighbour' talked to Duncan Law of Transition Brixton and celebrated the fact that actually talking to your neighbours is the new Gucci. On Thursday 6th September, a year to the day after its Official Unleashing, Transition Town Totnes celebrated its first birthday at the Royal Seven Stars Hotel in Totnes. The sell-out event was an opportunity to reflect upon the achievements of the previous year and to look ahead to where TTT might go next. It was a great example of John Croft's exhortation that Transition Initiatives should celebrate as often as possible.

The evening was a delightful celebration of the year, with good food, music, dancing and laughter. It was a fitting close to a quite extraordinary year, and a fitting beginning to the next one.

The first talk of the new programme was an evening called 'Redesigning our towns and cities for life beyond the car', which featured talks by Herbert Girardet and Peter Lipman. A dazzling array of ideas was put forward for how a town like Totnes could break its love affair with the car. This was followed by an Open Space day on transport, which came up with some good ideas.

As September draws to a close, the first steps are being taken towards the creation of an Energy Descent Action Plan for the town and TTT is a key player in a coalition being formed to develop the site of Dairy Crest, previously the town's foremost employer, which has now closed with the loss of 160 jobs. The Totnes Pound continues to be widely used, we are in discussions with a local developer about 'transitioning' a major housing development in the town, TTT made a major submission to the Local Development Framework when it came up for review, and the Totnes Renewable Energy Society launched in early November. Transition Tales will be working with a larger group of kids at KEVICC School and is expanding into other areas, such as working with the local art college getting them coming up with some Transition Tales. The programme for 2008 is focusing much more on practical projects, and we are also running a five-minute film competition on the theme of 'Totnes in 2030'.

To celebrate one extraordinary year since its Official Unleashing, you are invited to

Transition Town Totnes's

First Birthday Party!

An Evening of **Music**, Reflection, **Comedy**, Conversation, **Poetry** & **Dancing**.

with **Matt Harvey, Bert Miller** and the **Animal Folk, Rooh Star, Ollie** and **Kirsty**, and Breton dance band **Poisson Rouge**.

Also including our **Ravishing Raffle, Canny Competitions** and other **fabulous** surprises.

The **Seven Stars Hotel**, Totnes.
Thursday September 6th 2007. 8pm - 11.30pm.
Tickets in Advance from Greenfibres and Totnes Pet & Garden. Raffle tickets available at Red Wizard Internet Café, Riverford Goes to Town and Harlequin Bookshop.

Totnes Pounds Accepted.

TOTNES

TTT's first birthday cake(s) – note marzipan replica of Totnes Pounds!

Some practical manifestations of the Totnes Transition

"You can't speak about anti-globali-sation unless you can demonstrate that viable local systems are do-able, positive and that communities will back them . . . and that life improves for people as a result.

Transition Town Totnes is creating models for people. It's creating visi-ble demonstrations of how people can live in a satisfactory way in a sustainable manner, with identifi-able principles and an identifiable way of doing things that will make great change on the impacts we make on the Earth. We can't pro-ceed without those. If we don't have these kinds of models then we don't have anything, because then it's just fantasy. Then it's just us saying 'let's do this' and everyone else saying 'yeah, but how?'. If somebody is showing how, then suddenly you have a picture of how, and then you can set out on that path, try to emu-late it. So it is absolutely crucial that these things proceed."

(continued on facing page)

What follows is merely a taster of the diversity of initiatives emerging in Totnes since the Unleashing. By the time you read this there will certainly be more, but it will hopefully give you a deeper insight into the diversity of initiatives coming forward.

1. Oil Vulnerability Auditing

Oil Vulnerability Auditing is a very powerful new tool for engaging businesses in Transition work, using a tool which has an obvious rele-vance and which appeals to their bottom line interests. It is naïve to expect any more than a handful of ethically driven businesses to begin to make major changes to their business prac-tices in the absence of a calamity to force them to do so, as most businesses are working flat out just to keep their heads above water. At the same time, a Transition Initiative that fails to really engage its business community is miss-ing something vital, and has much reduced potential for success.

Simon Snowden of Liverpool University has been looking at this, and is developing the tool of Oil Vulnerability Auditing (OVA).[11] In essence, it is a method for auditing the various processes a business uses, and where it utilises oil, whether directly as fuel, as lubricants, in transportation, processing, packaging and so on. It allows the person conducting the Audit to build up an accu-rate picture of where oil is used, and then to explore, by pushing up the price of that oil, where the business's vulnerabilities lie. At $100 a barrel? $120? $150 a barrel? Which parts of the business's operations become unviable first? Is it the degree of dependence on transportation for the goods that they sell, prompting them to explore more local sourcing, or is it the energy-intensiveness of their processing?

OVA is a risk-assessment tool. It looks to the bottom line, and requires no allegiance to the peak oil/climate change arguments. It comes as no news to people, especially to those running businesses, that oil prices are rising, are not going to go down again, and that their operations are increasingly affected by them. Totnes is the first community to run a training in OVA, and to begin to use it with local busi-nesses. This tool has wide application, and as well as being used to assess businesses, it might also be possible to use it to assess Council planning policies and the vulnerability of new developments, such as when new sites are being selected for housing developments.

2. Skilling Up for Powerdown

Skilling Up for Powerdown is a ten-week evening class which aims to train people to be the field workers for the project. It is run in the centre of town once a week and has, thus far, been very well attended, with many of those who complete the course going on to become very active in TTT. The programme for the course runs roughly like this:

Week 1: Peak oil and climate change
A grounding in the two issues that will under-pin the rest of the course.

Skilling Up for Powerdown students discovering each others' skills and needs, using a 'speed dating' approach.

Week 2: Permaculture principles

A crash course in permaculture principles and how the Transition approach is based on them.

Week 3: Food

How oil-dependent the current food supply system is, and what an oil-free system might look like.

Week 4: Energy

How we use it now, how we can reduce the amount we use, how a powered-down settlement might derive its energy.

Week 5: Building

The distinction between 'green' and 'natural' building, the potential renewal in the use of local building materials, slides of examples, and a practical cob-mixing cookery class.

Week 6: Waste, water and toilets

The wastefulness of our current system, composting loos, town-scale urine harvesting, how much rainwater can you collect from your roof, water conservation.

Week 7: Economics

What money is, and where it goes, why we need local money systems, LETS, Timebanks, printed currencies, the Totnes Pound.

Week 8: Trees and woodlands

Forest gardens, the need for urban trees, nuts, fruits, forest garden design, my ten favourite trees.

Week 9: The psychology of change

Why do we not change, and how can we help people to change? Where does change come from, and how does it get blocked? This is also where we do the Twelve Steps exercise described on page 90.

Week 10: Pulling it all together

How does what we have covered on the course inform the Transition Town Totnes process and vice versa?[12]

The course also has lots of games and activities, slides and groupwork. It is backed up by an extensive set of online resources, links and reading lists which deepen the students' potential for learning. I plan to develop a *Skilling Up Teacher's Guide* in the near future.

3. Totnes, the nut tree capital of Britain

We are fortunate in Totnes to have, on our doorstep, the Agroforestry Research Trust,[13] one of the world's leading authorities on productive varieties of nuts and other useful trees. Over the past fifteen years they have trialled and bred a number of varieties of nut tree which show promise in the UK's temperate climate. These include sweet chestnuts, walnuts,

"On a larger scale, there will need to be new regional systems; I mean, trains run from here to the other side of the country so there have to be viable connections, so we have to operate also on a regional level – it can't only the local. Probably it even has to be global in certain contexts; we need to have some global systems for dealing with global catastrophic problems. . . .

There are plenty of reasons why there are common interests on a global and international scale, so you need some kind of systems in place for that but you don't want the economics part of that to be the dominant part of it, you want the collaboration and the consciousness part of it to be part of the story. The crucial thing is whether local systems can be set up that actually work great, and if they are, then there can be no real resistance to moving in that direction, so that's why the spirit is very important."

– Jerry Mander, interviewed in Totnes, March 2007

heartnuts, hazelnuts, pecans, almonds, butternuts and buartnuts (a cross between heartnuts and butternuts – a terrible name, but great nuts).

I often argue that it is only in the past 40 years that easy oil has afforded us the luxury of being able to design totally useless landscapes. Prior to that, our townscapes were scattered with allotments, gardens, orchards and so on. The epitome of where we have got to now is the landscaping around a newly created Business Park. Trees chosen for their aesthetic rather than their practical value sit above 'low-maintenance ground-cover shrubs' – plants bred specifically to be entirely useless. Why plant a flowering cherry when you can plant a fruiting cherry, which by necessity has to flower in order to then be able to fruit? This sanitised, 'tidy', low-maintenance concept of what our urban landscape should look like is both dysfunctional and an utterly wasted opportunity to create resilience.

I remember many years ago when I lived in Italy, visiting a nearby village called Santa Luce in the autumn. The main street was on a hill, and lined on both sides with aged, venerable sweet chestnut trees. The road was, although not quite ankle-deep, awash with large ripe chestnuts. However, no-one was collecting them, or appearing to show any interest in them. When I asked why, I was told that they were considered to be 'war food', something they had had to eat rather than chose to. That image of a main street awash with nuts has always stayed with me. Walnuts and chestnuts can produce as much carbohydrate and protein per acre as most grains can, and are constantly locking up carbon, offering shelter, and moderating urban temperatures.

The idea is to run a process whereby we get as many nut- and fruit-bearing trees into the urban fabric of Totnes as possible. Walnuts aren't in fact the first choice for this, as one would also need a fairly drastic squirrel-culling process in order to ever see any of them, whereas sweet chestnuts and heartnuts offer a better chance of seeing productive harvests.

We began with the Mayor of Totnes hosting a walnut and almond tree planting in the middle of the town. We then ran a design day, where people were given maps and went off round the town identifying places where more trees could go. This scheme has the support of the Council Tree Officer. In December 2007 another thirty trees were planted and more again will be planted in February 2008.

4. Totnes Local Food Directory

Local food directories are not a new idea, and are by no means unique to Totnes. Many Local Agenda 21 processes led to the creation of such directories, and quite a few of them now exist: some for towns, some for areas, and others for regions. They are, however, still an extremely useful strategy and a vital tool in the relocalisation process. Although one exists for the wider South Hams area, Totnes has never had one. The group who created this emerged from the Skilling Up for Powerdown course, and they set about visiting each food-related business in the town and creating a database of what they grew, produced or sold. They made a conscious decision from the outset not to include the supermarkets, and also to try to produce the Directory without advertising. A small amount of funding for the printing was secured, and various local writers and artists contributed pieces. The Directory is given away free in food outlets across the town, and has been very well received.

The cover of the Totnes Local Food Directory.

5. The Totnes Pound

One of the key challenges of any relocalisation process is that of economics; the fact that so much money, and the potential it represents to make things happen, pours out of the community in much the same way as water pours from a leaky bucket. The idea for the Totnes Pound came about at a talk I attended at Schumacher College by the alternative economist Bernard Lietaer. He said two things that were very striking.

The first was that relocalisation is only possible with both national currency and a local currency, and the second was that any such currency must be designed from the outset so that businesses will use it. This for me addressed very usefully the reason why LETS schemes (Local Exchange Trading Systems) are not really up to the job of economic relocalisation. While they have an essential role to play, they tend to have a limited lifespan, and rarely make the step into being of much use to local businesses. Given the scale of the challenge presented by peak oil, and the degree of urgency in the rebuilding of local resilient infrastructure, likened by some to a 'wartime mobilisation', we felt Totnes needed more than LETS.

Of all the initiatives that have sprung up in Totnes since the Unleashing, the Totnes Pound has generated the most press coverage and has achieved the highest profile within the town. As a project, it is a good example of just deciding to have a go at something rather than convincing yourself that it is impossible, or somehow that it will be stopped.

Following Bernard Lietaer's talk, we began to conceive of a printed currency which could only be used locally. We asked around about the legality of printing your own money, and were told that as long as it didn't pretend to be sterling, it was fine. We wondered why, if it was that easy, everyone wasn't doing it? It is a question to which we still haven't found a good answer. The most inspiring models we came across are the Berkshares currency [14] in the Southern Berkshires region of Massachusetts, and the Salt Spring Dollar,[15] produced on Salt Spring Island near Vancouver.

These are both printed currencies, with a full spectrum of notes ($1, $5, $10, $20 and $50). Each note features on one side a picture of someone felt to have been pivotal in the community's formation and development, and on the other side, the work of local artists. Both are purchased into circulation. In the case of the Berkshares currency, the notes are worth 10% more than the national currency: you can buy

"Maybe all this [Transition Initiatives] sounds a bit goody-two-shoes to you – a bit ecofreaky – but what's wrong with that? We've been here before. When our food supplies were threatened in the last war the Government urged us to dig for victory . . . and we did. Never in our history have we had a more healthy diet. And the fact is that people are responding to Transition schemes. They're packing town and village halls around the country to support them.

You don't believe there's any need even to think about this sort of thing? You reckon this latest oil crisis is just another scare and the danger of global warming is being exaggerated? Well maybe you're right. I hope you are. But if you're wrong, doesn't it make sense to think local rather than rely on politicians at national and world level to get us out of the mess they've helped create?"

– John Humphrys, The Sunday Mirror, November 25th, 2007

Photo by Noel Longhurst

Spending a Totnes Pound in the local market.

The experience from the Berkshares currency is that about 25% of the notes put into circulation are taken away by visitors as souvenirs, meaning that the money used to buy them stays behind in the community.

One side of the first notes we designed carried a facsimile of an 1810 pound note, printed at a time when Totnes Banks were allowed to issue their own money. The other side told people the eighteen local shops where the notes could be spent, and also had a panel of boxes which people were asked to write in each time they changed hands so we could observe where the notes went and how often they were used (see below). We worded the note in such a way as to provide an extra get out in case we had got this all horribly wrong and had overlooked something major: "This is a private currency which you may, by mutual consent, wish to treat as being roughly equivalent to One Pound." Not being worth a

$22 worth of Berkshares for US $20, and a Berkshare can be used in local shops as being equivalent to $1. This 'subsidy' to the local is very important, meaning that these currencies are attractive to the buyer and so become really key instruments for supporting local businesses above chain stores and multinationals.

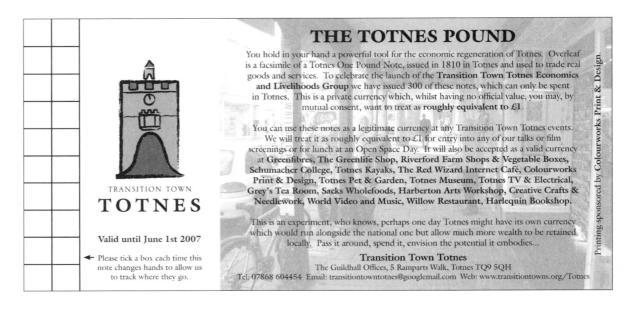

TRANSITION TOWN
TOTNES

Valid until June 1st 2007

← Please tick a box each time this note changes hands to allow us to track where they go.

THE TOTNES POUND

You hold in your hand a powerful tool for the economic regeneration of Totnes. Overleaf is a facsimile of a Totnes One Pound Note, issued in 1810 in Totnes and used to trade real goods and services. To celebrate the launch of the **Transition Town Totnes Economics and Livelihoods Group** we have issued 300 of these notes, which can only be spent in Totnes. This is a private currency which, whilst having no official value, you may, by mutual consent, want to treat as **roughly equivalent to £1**.

You can use these notes as a legitimate currency at any Transition Town Totnes events. We will treat it as roughly equivalent to £1 for entry into any of our talks or film screenings or for lunch at an Open Space Day. It will also be accepted as a valid currency at **Greenfibres, The Greenlife Shop, Riverford Farm Shops & Vegetable Boxes, Schumacher College, Totnes Kayaks, The Red Wizard Internet Café, Colourworks Print & Design, Totnes Pet & Garden, Totnes Museum, Totnes TV & Electrical, Grey's Tea Room, Sacks Wholefoods, Harberton Arts Workshop, Creative Crafts & Needlework, World Video and Music, Willow Restaurant, Harlequin Bookshop.**

This is an experiment, who knows, perhaps one day Totnes might have its own currency which would run alongside the national one but allow much more wealth to be retained locally. Pass it around, spend it, envision the potential it embodies...

Transition Town Totnes
The Guildhall Offices, 5 Ramparts Walk, Totnes TQ9 5QH
Tel: 07868 604454 Email: transitiontowntotnes@googlemail.com Web: www.transitiontowns.org/Totnes

Printing sponsored by Colourworks Print & Design.

specific amount we hoped would lessen any tax implications for the shops. A local printer offered to sponsor the printing and so we ran off 300 copies, in full colour.

We launched the Totnes Pound on Wednesday 7th March 2007 at an evening called 'Local Money, Local Skills, Local People' which featured a talk by economist Molly Scott Cato and which was the launch of our Economics and Livelihoods group. One of the motivators to the launch of the Pound was the question of how to get people along to an evening about economics. It is a subject that rarely prises people away from their sofas, so we thought, why not, we'll pay them to come! We gave them a Totnes Pound on the door.

People were encouraged to spend Totnes Pounds and to circulate them in the town, but I suspect that at least half of them were pinned on to fridge doors and dusted adoringly every morning, as they are rather collectable and gorgeous. Eighteen local shops were ready to take them, ranging from food shops to printers, kayak sellers and tea shops. Each shop had a sign in the window which read 'Totnes Pounds Accepted Here', followed by a list of the shops that would accept them, and then suggesting that people ask for them in their change.

At the beginning, some shops were a bit worried that they would end up with an excess of Totnes Pounds in the till and would finish out of pocket. We allayed their fears at the start by presenting the scheme as, in part, a promotional exercise for them, having the name of their business printed for free on things that would be passed around by hundreds of people. It turned out, on closer inspection, that the shops where the backlogs were occurring were the ones who didn't have the 'Please Ask for Totnes Pounds in Your Change' sign on the till. Once we gave them these signs, they found that

at the moment when people were stood at the till, they tended to ask about the Pounds. By the end of the scheme, shops were asked for notes far more times than they were able to give them out as change, and there were lots of disappointed tourists who had heard of the scheme and wanted to take one home.

One tourist was stopped and asked by some members of the Totnes Pound team who were making a short film [16] if she had ever heard of the Totnes Pound. She replied: "Is this the Pound where they keep people who were naughty?" Some of the shops who were involved started coming forward with lovely Pound-related stories. One local bookshop owner said that a woman had come in, bought a book using her Pound, gone out, and come back two minutes later to buy another book so she could get her Pound back in the change. Some of the traders on the market started accepting the Pound.

This first release of Totnes Pounds was always seen as being a pilot. It was due to run until 1st June, so as to allow us to assess how it went, and also so we knew it was going to end, in the event of it all going horribly wrong for some unforeseen reason. The week before it was due to end, we went around and interviewed all of the shops who had taken part. The main findings from that were that people felt there were not enough notes in circulation, that they were too big (they were the size of a standard compliments slip), and that the shopkeepers, customers and visitors to the town all loved them.

A Totnes Pound team was then formed with the aim of developing a larger scheme which incorporated the lessons learned from the first notes. [17] We looked around at the ways other currencies had been bought into circulation, but discussion with the team behind the Berkshares currency led us to think that 10%

was too high a margin, and we opted instead for 5%, so as to still provide an incentive for their use but not put too much of a burden onto the local shops, the 5% being just enough to encourage people to believe in them as being worth £1. The notes were designed to be smaller, to be on a more durable paper, and a wider range of shops were invited to participate.

Five shops throughout the town were invited to be issuers of the notes, so any member of the public could buy £T10 for STG£9.50. The options for the shops are to:

- spend the notes themselves in any of the other businesses in the scheme
- give them out as change (the preferred option)
- pay their staff partly in Totnes Pounds
- or exchange them back with TTT for 95p each

The idea was that the 5% 'devaluation' on exchange would provide an incentive to keep them in circulation. We think that the potential of this scheme is huge. Might charitable organisations or community development groups who want to invest money in the community do it with Totnes Pounds, thereby increasing the potential amount of work their money can do by 5%? To what extent will they be used by people as well as businesses? Will they be used for babysitting and helping out on allotments? Time will tell.

As a visible manifestation of TTT, it has been a huge success. Indeed in an informal poll in the town, more people had heard of the Pound than had heard of TTT! What these notes do is tell a new story about money and our relationship with it.

The third phase of the Totnes Pound was launched in January 2008. I think it is the most beautiful of the three notes so far. For the new note, as part of the ongoing experiment, the exchange rate is one for one, which we hope will enable some of the shops with the biggest turnovers not to have to struggle to give them out, and will put the onus on TTT to encourage them rather than the shop. I think people are thrilled by the potential that these notes embody. Where it all goes remains to be seen.

6. Transition Tales

Transition Tales is a broad-based initiative which aims to use storytelling and imagination work as a way of looking forward to a lower-energy future, as part of the visioning work described in

Students at King Edward VI Community College in Totnes discussing what they would miss and what they would look forward to in an oil-scarce world, as part of a Transition Tales workshop.

Part Two, pages 94-121. It is designed to engage both adults and children. The idea is to work with the imagination, to help people actually imagine themselves in the future. With adults, we have begun to invite them to create stories from the future, some examples of which have appeared earlier in this book (see Chapter 7).

This work offers a great opportunity to engage artists, writers, poets and creative thinkers in looking forward. A recent Transition Tales Storytelling Day in Totnes began with a local writer giving participants a crash course in the creative writing process. The stories generated could either be used to create a *Futures Gazette*, a newspaper from the future, or to illustrate the Energy Descent Plan.

Working with children requires a different approach. We have created a programme of workshops which introduce the concept of peak oil to children and get them to create positive visions of a future as oil becomes more scarce. We have been piloting this work with the local secondary school, King Edward VI Community College, with very promising results, and in April 2008 are running workshops for the entire Year 7. We aim to develop the idea further across other schools (you can see some of what we are developing on page 118). We'd like to produce a *Futures* newspaper, compiling some of these stories into a newspaper format.

7. TTT Home Groups

Change is difficult to initiate, but even harder to sustain; to use the phraseology of the Stages of Change model discussed in Chapter 6, once we have decided to take action, we need to be able to maintain that change. We often feel inspired to do something, but everyday life has a habit of taking over and our plans to do things can often end up on the back burner. A number of movements in recent history have explored how to make change 'stick', in particular the women's movement in the 1970s. One of the tools they experimented with and evolved was 'consciousness-raising groups', in which people offered each other support while implementing difficult changes in their lives. The same model has been adapted and adopted by the peace movement and also, more recently, by the men's movement.

Home Groups are an initiative of the Heart and Soul group of TTT, who are exploring the psychology of change. They are concerned that people exposed to issues of peak oil and climate change can find themselves overwhelmed by the scale and the implications of the information they are being presented with, and struggling to make meaningful change in their lives. Creating Home Groups, which they define as "small, close groups in which people get to know each other well, with a shared intention", can offer a degree of support and mutual reinforcement which environmentalists rarely feel. These groups will allow people to share their excitement, their skills and resources and their energy for making practical change.

In TTT, some of these groups have arisen as a continuance of the Skilling Up for Powerdown evening class, which runs in ten-week blocks. They offer a dynamic way for people to sustain the energy generated on the course and to support each other in implementing the changes inspired by the course.

The viral spread of the Transition concept

Running alongside the unfolding of the Transition process in Totnes and Kinsale was a rapid take-up of the idea. Within weeks of the Unleashing other places were getting in touch to ask what we were doing, and soon this had become a viral phenomenon. Some of the key events along the way included:

- 'One Planet Agriculture', the January 2007 Soil Association conference in Cardiff which was focused on peak oil and relocalisation, and proved to be their most popular ever[1]

- a talk in Lampeter in west Wales, organised by the West Wales Soil Association, which raised the idea of Transition Town Lampeter[2] and was attended by over 400 people. It has since gone on to be the catalyst for a lot of the towns surrounding Lampeter to start their own Transition processes, with Lampeter as the hub

- lots of media coverage, including ITN News, *The Guardian*, a whole programme on BBC Wales and also an excellent piece on BBC Radio Scotland.[3]

- the Official Unleashing of Transition Town Lewes, which took place in the Town Hall, was attended by about 450 people and which as well as Unleashing the Lewes process, triggered other similar projects in surrounding communities

- a meeting with Prince Charles at his Food and Farming Summer School at Highgrove and giving him a Totnes Pound.

Since Transition Town Totnes Unleashed, the Transition model has been taken up by communities all over the UK and increasingly beyond. The demand became such that we set up an organisation called The Transition Network to most effectively support them (see Appendix 5). The Transition model is a simple one, and each community that gets involved is contributing valuable research as to what works and what doesn't, and how the model needs to be adapted for different scales, settings and cultures.

What follows is a snapshot of seven Transition Initiatives on a range of scales, in the order in which they emerged, to give you a feel for how this idea is being interpreted in different places. It is important to remember that at the time of writing, the most advanced of these has only been going for just over eighteen months.

Transition Penwith

Penwith is not a town or a city but a peninsula, the westernmost district on the UK mainland. It has a population of 63,000, but given the land area of Penwith, this means that overall the density is only 2.07 people per hectare. Most of those live in urban centres, but there are many small scattered rural communities too.

Transition Penwith (TP) came about when

Jennifer Gray attended a small informal think-tank meeting with myself and a few others to look at various aspects of the Transition process. This was prior to the Unleashing of TTT, and Jennifer felt that what seemed to be emerging in Totnes could also work in Penwith. It then transpired that Richard Heinberg would be in the country soon, and the idea was born that Transition Penwith would start with what Jennifer called a 'Big Bang', rather than with the awareness-raising stage. In the run up to Heinberg's talk a lot of time and energy went into publicity and into inviting many people from all aspects of the community and a range of organisations. In the end over 400 people turned up, including the area's MP and the Town Mayors from Penzance, St Just, Hayle and St Ives.

The event created a huge wave of interest. Jennifer talks of the buzz it created, about how walking down the street for a couple of weeks afterwards everyone was talking about the evening, about Transition Penwith and about the implications of the talk. TP's first programme was modelled quite closely on the TTT programme, and also adopted the same design template. It included talks by David Fleming and myself among others, and also film screenings such as *The End of Suburbia* and *The Power of Community*, all facilitated in the way set out on page 154. Working groups began to emerge spontaneously, and a large number of partnerships were formed with a wide range of groups. These partnerships were important in not duplicating work already being done elsewhere and also in being part of the strategic framework in the area. Transition Penwith have since organised many events with these partnership organisations.

Some of the practical projects that have emerged from the TP working groups include:

- A study looking at energy usage and requirements for the area as a whole

- Transition Energy Tours – tours of local renewable energy installations

- 'Chutney Rules' – home-based reskilling workshops, doing things like building a worm bin, installing a solar panel, making a raised vegetable bed, and so on

- workshops on a range of skills such as compost toilet building

- a 20-week course in renewable energy skills

- a Transition Roadshow created by the Education Group to take around village halls and schools in the more outlying areas of Penwith

Transition Penwith has acted as a catalyst, along with Transition Falmouth, for the many emerging Transition Initiatives in Cornwall. They are endorsed by the local District Council, who have come on board as partners and who provide resources such as meeting spaces and audio-visual equipment.
www.transitionpenwith.com

Transition Falmouth

At the Permaculture Convergence in Dorset in 2006, Lorely Lloyd heard a presentation I gave about peak oil and the Transition concept, and had her 'peak oil moment'. She went home to Falmouth, where she was a Town Councillor, and thought about it all for three weeks until at a conservation event in the

"What makes the Transition movement so attractive is that its message is simple and unfailingly positive. It is place-specific in that it addresses the concerns of individual communities. It looks at what can be done rather than the possibility that it may be too late to save the planet from runaway climate change or global oil crises resulting in systems failure. . . . It's also a sign of the times and an answer for anyone who saw An Inconvenient Truth *and wondered, 'What can I do?'. The Transition movement has harnessed the collective call to action and is a glue that is mending the torn fabric of our communities."*

– Cliona O'Conaill (2007), 'Carbon Descent: the Transition movement is a local solution to a global crisis', *Resurgence* No. 244, September/October 2007

town she began headhunting people to form a steering group for what would become Transition Falmouth.

In October I went to Falmouth and gave a talk to the fledgling group in the Town Council chambers. Those attending ranged from a 13-year-old boy to the Town Clerk and the Town Centre Manager, and the talk gave them an overview of the what, why and how of the Transition approach. After the talk there were meetings every two weeks, talks, films or discussions. There was also a Powerdown Christmas party, and a number of working groups began to form.

One of the key areas of Transition Falmouth's (TF) work has been in building partnerships. They have made strategic partnerships with Friends of the Earth, the Town Centre Forum, the Town Council, Falmouth Green Centre and many other groups. One of the key achievements from their partnership-forging has been the development of an integrated transport package for the town centre, including the pedestrianisation of the High Street. The Transition approach of embedding peak oil and climate change and looking for common ground managed to draw together previously disparate groups to produce important local action.

Some of the practical projects that have emerged from TF include:

- 'Darnit' – an arts and crafts group which brings people together to learn/share knitting, sewing and other craft skills
- The beginnings of a move to make Falmouth plastic-bag free

- A food group, whose first event was Seedy Saturday – a seed-exchange day
- Supporting Falmouth School to become an 'Eco-School'

With their local Arts Centre showing the film *A Crude Awakening* and Falmouth University College starting to do more talks on issues relating to Transition, TF is moving away from organising programmes of events and is able to focus more on practical projects. There are also plans, in partnership with the Town Council, the Allotment Association and Falmouth Green Centre to create a Sustainability Centre, which would host education, reskilling and community development activities. They are also applying for funding for edible pathways throughout the area. Reflecting on the first year of Transition Falmouth, Lorely says:

"As a tourism and dock-based town with a high elderly and student population we are finding it quite a challenge to engage the majority of people, but Transition is an excellent peg for all kinds of groups and individuals to hang their

Transition Falmouth's Seedy Saturday event.

hopes for the future. We are steadily forming a diverse and committed team who are establishing projects and partnerships to suit all who wish to join us in working towards our resilient community. Working with other local Transition Initiatives is proving excellent."

Transition Town Lewes

Lewes in Sussex was one of the first Transition Initiatives to get underway. They have modelled their approach very much on that of Totnes, and have also inspired other emerging Initiatives in the area. Transition Town Lewes (TTL) began when Adrienne Campbell heard from a friend about the Totnes Unleashing. Inspired, she and others set up a steering group and TTL was born.

The first programme of events they did was called 'Energy Transition Comes to Lewes', and it included talks by Jeremy Leggett and Caroline Lucas, as well as facilitated screenings of *The End of Suburbia* and *The Power of Community*. This led on to the Official Unleashing (making Lewes the third Initiative to do so) on April 24th 2007. Over 400 people came, the Town Council finished their regular meeting early so they could attend, and the venue was also filled with stalls of other local groups.

After an introduction from the Mayor, I spoke, followed by Dr Chris Johnstone. The event was very energising, with lots of enthusiasm about the future. Following the Unleashing, Transition Town Lewes has been very busy. They have run Open Space days (see pp.168-9) on food, housing, energy and their Energy Descent Action Plan, as well as a full programme of talks, film screenings and events, including Slow Food feasts and a very busy Great Reskilling programme.

Not content with the above, TTL has also catalysed a number of interesting projects. These include:

- Work in local schools
- 'Heat your greens', a local solar panel buyers' club
- 'Grow your own vegetables', a series of practical workshops
- The Lewes Community Bag – a cloth shopping bag, printed with 'Love Lewes, Shop Local' on one side, and the logos of twenty local shops on the other, produced as a limited edition of 1,000.

Transition Town Lewes running a workshop on peak oil and climate change in a local school.

Photo © Adrienne Campbell

There are a number of other schemes in the pipeline too. These include the Lewes Pound, a car club, something called the 'Book Crossing of TT Books',[4] which is about leaving books in public places to read and then pass on, the mapping of existing orchards and creation of a forest garden, and the creation of OVESCo (a community owned energy company for the Ouse Valley Bioregion).

Although it wasn't the very first Transition Initiative, TTL has a few 'firsts' under its belt. One is that it is the first to really implement Step 1 of the Twelve Steps of Transition, 'Form a Steering Group and design its demise from the outset' (p.148). Having formed a dedicated group of people who had navigated the project through the first few steps, it was decided to wrap up the group and hand the project over to representatives of each of the Working Groups. This was by all accounts not an easy process: for those who had been there from the start it required a great deal of letting go and trust that it would work.

However, on reflection, it is seen to have been a useful process. Adrienne describes it thus:

"A dozen people willing to form working groups had emerged from the Unleashing and the core group started to discuss our demise. Fortunately, some professional 'facilitators of change' turned up who invited the group contacts to create a new set of aims and principles. If applied with responsibility and openness, these empower the working groups to work autonomously. Our demise was a painful process for me, a kind of paradigm shift, that required a good deal of trust. Some of the original core group has left TTL, and some of the groups are struggling, but I feel the overhaul was necessary to allow Transition Town Lewes to really fly.

There's always been a creative tension between people who are more motivated by either peak oil or climate change issues. There's also a dynamic between the campaigning types – for change now – and the Energy Descent planning/permaculture approach. We're having an ongoing discussion about how divergent our attitudes can be, and whether even an assumption of 'life with less oil' is a given. And if not, what is the transition we are making? The background global context of continuous growth and erosion, plus the lack of a language to describe the inevitable paradigm shift, makes this whole community-building enquiry rich, scary, inspiring and surreal all at the same time."

www.transitiontowns.org/Lewes

Transition Ottery St Mary

This initiative began with Sara Drew attending a talk by Peter Russell in January 2007, presented as part of the Transition Town Totnes programme. "I felt the energy at the event, and thought 'wow!'. Here at last was a vehicle that was positive rather than negative, and wasn't about falling into a pit of despair!" She went home and wrote a letter to the local paper, which ran as follows:

"I live in Ottery St Mary and have spent a lot of time reflecting on the state of the planet, and the coming crisis of resources – especially oil. For instance, we consume six barrels of oil for every one we produce! Put that with carbon emissions, melting ice caps and runaway climate change and it is easy to feel we are on the edge of an abyss which is going to claim us all. The system is completely unsustainable.

However, I recently attended a talk in Totnes where the community is coming together to develop practical ways of doing something about all of this in their own backyard. Their project is called Transition Town Totnes and is all about developing highly local ways of reducing energy consumption, building a better community and taking charge of their own lives to cope successfully in a world after oil – a Post Carbon Community. Their ideas include food-growing groups, transport sharing and energy efficiency. There are groups like this getting together all over the planet. I am sure we can build a Post-Carbon Ottery which will improve the quality of all our lives and help us get through the coming years – together. If you are interested in getting together to make it happen please contact me (details provided)."

A number of people responded, and the core group of Transition Ottery formed. It has twelve members, and is working on the principle of not rushing, but taking the time to really put down deep roots and establish connections in the community. They have run a series of events, attracting over 100 people to a screening of *An Inconvenient Truth*, running a Family Green Day (a fête with local food, solar energy, and a computer where one could measure one's carbon footprint, among other things).

They are designing a lot of their programme for families with young children, and are working with the local Later Life Forum, to draw in older members of the community. Screenings of *The End of Suburbia* and *The Power of Community* are planned, and they plan to Unleash in summer 2008. They see their evolution as slow but steady, and they have built strong networks with existing groups, such as local composting groups and local food growers. *www.transitiontowns.org/Ottery-St-Mary*

Transition Bristol

The genesis of Transition Bristol (TB) came out of a permaculture design course that was run in Bristol in 2006. In discussions that emerged from the course, people explored what projects they might like to get involved in, and the idea was raised of doing an Energy Descent Action Plan for the whole city. Although the idea seemed pretty daunting, it took root. Some time later, there was a Permaculture Day School which brought a large group of people together for a day of workshops and practical skills training, after which a group formed (originally under the title 'Transition City Bristol').

No one really had an idea how the concept might work on such a scale, but all of those involved felt it important that they try. The first steps were forming the Steering Group and beginning the awareness-raising stage. The first high profile talk was by Dr Chris Johnstone, which made it on to the local BBC evening news.

In May 2007 I gave a talk at the Trinity Centre in Bristol, at the end of which people from the various parts of the city organised into their respective 'villages' and discussed how they would like to evolve this process at the local level. Bristol, like all cities, is formed from a collection of 'villages', each area having a distinct identity and this feels like the ideal scale for much of this work.

Photo by Tulane Blyth

Information board at Transition Bristol's Big Event.

Transition Bristol, as a result, now works on two levels. One is to go through the Transition process in order to work towards an Energy Descent Action Plan for the city as a whole, and the second is to network, inspire, train and support the 'village-scale' initiatives in the city. To support the villages, TB organises monthly meetings to enable them to share what works and what doesn't work. Alongside this, TB runs a programme of events and talks, a website and are promoting a 'virtual orchard' initiative, where fruit trees will be made available at cost across the city.

Working on the scale of 400,000 people is clearly different from working on the village and market town scale. According to Peter Lipman, one of the founders, the Twelve Steps of Transition were very useful in designing their strategy:

> "They helped us to slow down and deal with the pressure to do everything now on both of the levels we're working at. That pressure is inevitable, given the scale of the challenges and the feeling of incredible urgency – but if we don't resist it, we feel it could lead to us failing to open out the process to as wide an audience as possible as

well as burning us all out. They have also helped us when working with the local Transition districts, reminding us to encourage them to think about the importance of not dashing ahead too fast, and about how crucial it is to broaden your base of support and not assume that awareness is already in place."

Another of the things that has also been very useful has been the Criteria for becoming a Transition Initiative (see Appendix 5, p.221), which Transition Bristol has been using with the 'Village' initiatives, stressing that there are criteria that need to be in place which will make their success more likely. TB are planning their Unleashing for Spring 2008, and are increasingly in dialogue with some of the other emerging city-scale projects such as Nottingham, in order to share best practice.

In November 2007 Transition Bristol held 'The BIG Event', the largest Transition event yet to be held anywhere. The main speakers were Richard Heinberg, Jeremy Leggett, David Strahan, Dr Chris Johnstone and myself, but there was also a full programme of workshops and talks on a range of other subjects. Nearly

400 people attended and the day was a dynamic, positive and inspirational introduction to the whole concept of Transition, as well as giving people a number of tools for beginning or supporting their local Transition Initiatives. *www.transitionbristol.org*

Transition Town Brixton

Possibly the biggest task taken on by a Transition Initiative is Transition Town Brixton in London. Transition Town Brixton started life as the Lambeth Climate Action Group, which was formed after screenings of *An Inconvenient Truth* at the Ritzy Cinema in 2006. Early in 2007 member Duncan Law[5] came across the Transition model, and was seized by the idea of campaigning for something rather than against. He was also struck by the degree to which adding peak oil to the climate change debate strengthened and deepened both arguments.

Duncan suggested that the group reinvent itself as a Transition Initiative, which was unanimously approved. The group then began their awareness-raising stage, and thus far has held five film screenings and has hosted nearly twenty different speakers. It has also organised 'green' walks, visits to interesting projects and panel discussions. This has considerably raised the profile of the initiative, and generated national media coverage. The group has also formed strong links with the local Council.

Working groups are starting to form, but the group is still putting in place the infrastructure to enable this to happen most effectively. When asked how useful the Twelve Steps have been, Duncan told me: "They have been extremely useful, but unfortunately they arrived slightly too late! We had already formed our steering group for the Lambeth Climate Action Group,

which hadn't formed with the intention of becoming a TT or with its own demise being an option. We are still working on how we might apply the first Step, and certainly aren't working on them all at present. In many ways we are still in the Laying the Foundations and Awareness-raising stages." They had planned to have an Unleashing at the end of the summer of 2007, but felt that they still hadn't reached the critical mass required to guarantee an explosion, so it has been put back to Spring 2008.

Being still largely in the Awareness-raising stage there are few practical manifestations of the project, although some that are planned include a Green Map of Brixton and an urban gardening education project based on a gardening site which is already established. The Hyde Farm Climate Action Network has formed in a local estate, is growing food in front gardens and has held energy-saving workshops, film showings and fruit-picking events.

I asked Duncan what he saw as being the difference between applying the Transition model in Brixton and elsewhere. He told me:

"One of things that is so different about trying to implement this model in Brixton as opposed to a market town or village is that Brixton is a post-industrial phenomenon – it is not based on an ancient model of sustainability. There is no pre-oil infrastructure in place we can learn from. The effect of peak oil on somewhere like Brixton will be a huge shake-up. No one is looking squarely at the dependency of cities – how to feed them, and what will happen when unemployment rises and house prices fall catastrophically, when there is no more money in the grant pots, and when local government is squeezed. Local government is still making cuts. We are saying to them: invest now in this new infrastructure. We are also exploring how to make relationships outside the

city that will support the city, connections to farmers, as well as putting in place local infrastructure such as organic food hubs."

Applying the Transition approach to urban areas is one of the cutting edges of this work, and information-sharing between the various projects is important, so Brixton, Bristol, Nottingham, Brighton and the other urban projects draw a great deal of inspiration and encouragement from each others' efforts. *www.transitiontownbrixton.org*

Transition Forest of Dean

The Forest of Dean in Gloucestershire clearly presents a very different set of challenges from those of the city of Bristol. It covers about 200 square kilometres and is home to about 75,000 people.

The Forest is bounded by the River Wye (and Wales) to the west, and the River Severn to the east. The district's northern boundary is northwest of the city of Gloucester in the Vale of Leadon.

Transition Forest of Dean emerged with Sue and Andrew Clarke, who both worked as environmental consultants and were becoming frustrated at the lack of local action on climate change. They contacted their local council to ask them what they were doing about climate change, the answer to which was not much, because the community hadn't yet told them that they felt it to be an important issue. Around this time, Sue began to focus her work at a more local scale and in March 2007 she came across the Transition Network website. Seeing that the Inaugural Meeting of the Transition Network was coming up soon at Ruskin Mill near Nailsworth, she booked a place and headed over.

Inspired and fired up by the event, Sue returned home to initiate Transition Forest of Dean. She wanted to form a steering group which was not primarily made up of people from a 'green' background, but which represented the broader community. Sue describes the make-up of the group thus:

> "We have two with science/environmental/strategic consultancy backgrounds, three with a permaculture background, one Friends of the Earth, one Green Party, one from green business (a local wood co-op), one from a housing co-op, and another a key representative of 'Foresters' involved with preserving our local traditions and rights (very important to the community here). We also have two people who are basically concerned individuals who keep the rest of us grounded and are motivated enough to want to make a change."

Given the dispersed nature of the Forest, and

the fact that driving from one side to other takes about 40 minutes, the group quickly realised that they needed to go to the community, rather than expecting the community to come to it. They therefore developed a 'roadshow approach' to take round the main towns, showing films and facilitating discussions afterwards, some of their events focusing on peak oil and transition, and others on the possible impacts of climate change specific to the Forest of Dean area.

Their perspective is that although there are distinct towns within the Forest, most people see themselves as Foresters first, and as residents of their towns second. They, therefore, feel that working on the scale of the whole area is the most skilful approach.

At the beginning, they were concerned that people would feel that Transition Forest of Dean was going to do everything for everyone. Now, as word gets round, people are coming forward, more people are offering help and time. To date their audiences have tended to be mostly of those over 40 years of age, and so they are designing events and publicity in such a way as to have a broader appeal. They have also recently been invited to present to secondary schoolchildren.

Other events they have organised include a monthly social networking get-together, a quarterly local food feast, and stalls at local events. In terms of the Twelve Steps (p.148), they see themselves as being still in the first three, though the recent purchase of a number of Electrisave meters for loan to anyone who would like to monitor their energy usage is a visible action that will reap immediate benefits as people become more aware of their energy-usage habits and how they can reduce their consumption.

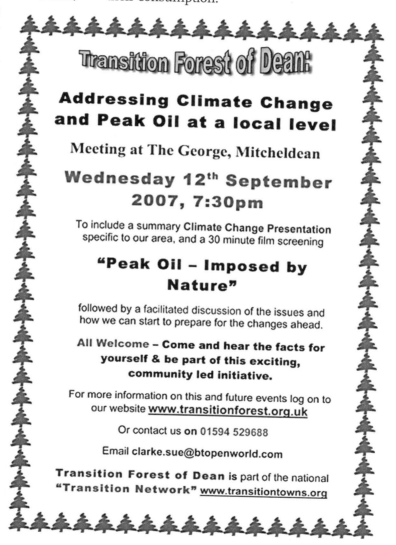

Transition Forest of Dean:

Addressing Climate Change and Peak Oil at a local level

Meeting at The George, Mitcheldean

Wednesday 12th September 2007, 7:30pm

To include a summary Climate Change Presentation specific to our area, and a 30 minute film screening

"Peak Oil – Imposed by Nature"

followed by a facilitated discussion of the issues and how we can start to prepare for the changes ahead.

All Welcome – Come and hear the facts for yourself & be part of this exciting, community led initiative.

For more information on this and future events log on to our website **www.transitionforest.org.uk**

Or contact us on **01594 529688**

Email **clarke.sue@btopenworld.com**

Transition Forest of Dean is part of the national **"Transition Network"** www.transitiontowns.org

"The hours when the mind is absorbed by beauty are the only hours when we really live."

– Richard Jefferies

Some closing thoughts

THE PARADOX OF OUR AGE

We have bigger houses but smaller families; more conveniences, but less time.

We have more degrees but less sense; more knowledge but less judgement; more experts, but more problems; more medicines but less healthiness.

We've been all the way to the Moon and back, but have trouble in crossing the street to meet our new neighbour.

We built more computers to hold more copies than ever, but have less real communication.

We have become long on quantity, but short on quality.

These are times of fast foods but slow digestion; tall men but short characters; steep profits but shallow relationships.

It's a time when there is much in the window

But nothing in the room.

— His Holiness the 14th Dalai Lama

Something about these profoundly challenging times we live in strikes me as being tremendously exciting. Clearly peak oil and climate change are potentially catastrophic challenges which can easily lead to our painting grim pictures of social breakdown and ecological collapse. It is easy to spend many sleepless nights worrying about them – many of us have. With a small shift in thinking though, such as that laid out in this book, we may find it hard to get to sleep due to our heads buzzing with possibilities, ideas and the sheer exhilaration of being part of a culture able to rethink and reinvent itself in an unprecedented way.

I believe that a lower-energy, more localised future, in which we move from being consumers to being producer/consumers, where food, energy and other essentials are locally produced, local economies are strengthened and we have learned to live more within our means is a step towards something extraordinary, not a step away from something inherently irreplaceable.

And what is it exactly that we are so fearful to move away from? The New Economics Foundation has shown that:

- Increased income stopped making us any happier sometime around 1961[6]
- When asked which decade from the 1950s onwards respondents would most like to have lived in, the 1960s emerged as most popular[7]
- 62% of people in the UK have jobs they find uninteresting or stressful
- 87% of Britons agree with the statement

'Society has become too materialistic, with too much emphasis on money and not enough on the things that really matter'

- The degree to which we trust each other has fallen by half since the 1950s.[8]

Also, with national indebtedness now at a record £1.2 trillion, it is clear that we are paying through the nose for something that fails to meet our fundamental human aspirations: happiness, security, time for relaxation, rewarding work and access to healthy food.

Those who are involved in Transition Initiatives (which might now include you if you have got this far in the book) are part of one of the biggest and most important research projects underway anywhere in the world. You are catalysing those around you to ask the questions that government still finds it very hard to ask, but that are essential to our collective survival. You are acknowledging that it is with us that real change begins, and that it is up to us whether we accept this responsibility or shy away from it. I hope you will find the tools in this book useful. What you do with them, how you assemble the pieces, is up to you. We are all making this up as we go along, creating the road as we travel along it.

Central to this book is the proposition that the future with less oil could be preferable to the present, if we are able to engage with enough imagination and creativity sufficiently in advance of the peak. I have mentioned this a few times through this book; I have done so because it is the Transition concept in a nutshell. What that engagement will look like in each community will emerge from that community itself. The idea that every village, town, city, hamlet, island or district could become a Transition Initiative may seem fanciful, yet it is entirely possible. It is an idea whose time has come. It is important to remind ourselves that this is not a process which has achieved nothing until it has completely powered-down its community; rather, what matters is the journey, the process, the coming together and doing it.

When I sat down to begin writing this book, oil cost around $70 a barrel. Now, in June 2008, the price has almost doubled, being close to $140. Previously optimistic forecasts are being continually revised downwards, with Fatih Birol, Chief Economist at the IEA, now saying: "We must leave oil before it leaves us." The Age of Cheap Energy is over. The sooner we can acknowledge that and can start to engage our collective creativity, the less likely it is that we will sink into despondency, blame-seeking and powerlessness, and the more likely it is that we will unleash the most extraordinary and historic transition. We are alive at a pivotal moment in human history.

While peak oil and climate change are undeniably profoundly challenging, also inherent within them is the potential for an economic, cultural and social renaissance the likes of which we have never seen. We will see a flourishing of local businesses, local skills and solutions, and a flowering of ingenuity and creativity. It is a Transition in which we will inevitably grow, and in which our evolution is a precondition for progress. Emerging at the other end, we will not be the same as we were; we will have become more humble, more connected to the natural world, fitter, leaner, more skilled and, ultimately, wiser.

We will emerge blinking into a new way of living, yet it will feel more comfortable and familiar than what we left behind. If we are to trade mobility, growth and affluence for something else, we need to be able to articulate something preferable and more nourishing to put in its place. I hope this book has inspired you to be a catalyst for exploring these new possibilities, in your life, your community and your world. May it keep you awake at night, but this time for all the right reasons.

"In the depth of winter I finally learned that there was in me an invincible summer."

– Albert Camus

"Another world is not only possible, she is on her way.
On a quiet day I can hear her breathing."
– Arundhati Roy

The 2008 Totnes Pound. In the near future, might such currencies be a central part of the daily lives of all of us?

Appendix 1

The Oil Depletion Questionnaire

as used in Transition Tales schools workshops

(In order to complete this worksheet you will need www.lastoilshock.com/map.html open in front of you).

Countries in red have already peaked in oil production, countries in green have not yet. Roll your mouse over the map to see when they peaked, their current production, and how much they were/will be producing at their peak. If you double-click on a country, you can zoom in; right-click to zoom out again. Use the map to answer the following questions.

1. Finland, Sweden and Greenland don't export oil. Why not?

_____

_____

_____

_____

_____

2. Which country produced more oil in 2005: China, India or Saudi Arabia?

_____

3. In which year did the United States peak in production?

_____

_____

_____

_____

4. UK oil and gas production peaked in 1999. We increasingly rely on gas from Norway and the Russian Federation. Why might this not be such a good idea?

_____

_____

_____

_____

_____

5. Which of the following countries has the most promising long-term oil supply – Kuwait, Iran, Iraq or Syria – and why?

_____

_____

_____

_____

_____

6. Countries such as China and Venezuela are increasingly standing up to the United States. What do you think might make them feel able to do so?

_____

_____

_____

_____

John Croft's 'Four stages that any project will go through'
by Naresh Giangrande of Transition Training

In Transition Town Totnes, we have found these four stages an extremely helpful tool and where possible incorporate them into our work. Early in 2007, John Croft, co-ordinator of the Gaia Foundation of Western Australia, visited Totnes, where he shared the following insights with us. He developed what he calls the 'Dragon Dreaming' method by examining hundreds of projects around the world, environmental, social justice, or community projects. He noticed that there are four stages of successful projects. He defines these stages as:

• dreaming or visioning

• planning

• doing

• celebrating and evaluating

He also noticed that 90% of projects failed to get beyond the first stage – dreaming. Of the 10% that did, another 90% failed to get beyond the second stage – planning. This means 1% of projects actually get off the ground and do what they set out to. In an effort to increase this low success rate he developed the Dragon Dreaming approach. His message is that each of these four steps is essential if your project is going to succeed. Leave out any one step and the project will invariably fail.

Let's take a closer look at these four in turn:

1. Dreaming
Each project is an encounter between the originator (any project, Croft argues, begins life as the dream of one person) and the environment, and between theory and practice. If you look at the project as a flow of energy, at the beginning the main flow will be from the 'dreamer' to the environment and the community around the dreamer. There will be very little coming back. In terms of theory and practice the project will also be almost all theory, as the practice has yet to be created. As the project progresses the environment around it responds. This is how you know you have a 'runner', a viable project. For instance, if you were developing a product and you had identified your target consumers and they indicated they would be queuing up to purchase your product, you would know you had a winner; the environment had responded appropriately. On the other hand, if there was a tepid response, then you would have to go back to the drawing-board. Maybe you misread your target audience; maybe the price was too high.

Either way there has to be a balance between the originator with his or her idea and the environment where the project is being launched at this stage before you go on to the next. Likewise the theory gradually takes on more material manifestations and moves from being merely a twinkle in the inventor's eye to something real and tangible. The sorts of questions asked at this stage are:

• What would happen if............?

• What would this sort of project look like?

• What do you think, does this sound like a good idea?

• Can you imagine our town with a ....?

- Don't you think that if people had access to .... it would be great?

The Dreaming Stage is the essential first stage, but in order to progress, it is vital to move into the next stage, the Planning Stage.

2. Planning

This is the stage in a project where an idea is worked up, where research and investigation take place. This is where 'the rubber really starts to meet the road'. The sorts of questions that need to be addressed in this phase of the project are:

- How do we make this happen?
- Who's going to design it?
- How many people in the team?
- What skills are we missing, or do we have?
- How might we finance it?

It is the beginning of an idea leaving the conceptual stage and engaging in the outside world. Are there already similar ideas out there? What can we learn from other successful models? What do we need in order for our idea to take practical form in the world?

3. Doing

This is where the idea is implemented. Something real and tangible is put in place. What was dreamed and planned becomes a reality. All the planning can now (if your dreaming and planning stages have been successfully completed) be put into practice. You have committed to something and are now seeing how this behaves in practice. The visioning and planning sessions in Stages 1 and 2 may raise questions which you do your best to answer, but the only real test is to actually do it. In Stage 3 you have signed your contracts, employed your workers, and installed the phone lines and your

baby has come to life. The theory is now practice, and with time and familiarity it becomes so second nature that you forget that it was only a theory not so long ago. This is the time of birth.

4. Celebration and Evaluation

The final stage is celebration and evaluation. This celebrates the success of the project and looks at the failures and difficulties before starting the cycle again. The next phase of the project (if there is one) is then dreamed, planned, executed, and celebrated. This process can go on and on as each phase of a project is brought into being. It is a fractal process: each stage also contains the other three parts of the process.

- Has the project reached your expectations?
- Has it matched targets you set, or your funders set you?
- Can it be, or should it be, done more or again?
- Has it reached all its target client group?
- Which phases of the project went well?
- Which phases were difficult?
- Was everyone properly recompensed?
- Was the project fun to work on?
- Was it hard work, adequately resourced, or did it rely on one person's unsustainable hard work and commitment?
- Did the people who worked on it grow personally, acquire new skills, or gain personal satisfaction from all their hard work?

When you have answered these questions and finished your honest evaluation, then the next phase can happen. Indeed, it will happen better and with more creativity as you now have had the experience. You can be more daring and take on something bigger; if that's what you want. It can go on and on in a spiralling cycle of success.

And now you can celebrate! You have done it. It is important on a group level and on a personal level to celebrate and integrate the new confidence and skills and awareness you have. One of the tasks will be to find an appropriate way to celebrate! If that sounds like a chore then you know which part of the project is weak! *Find out more at www.gaia.iinet.net.au.*

Appendix 3

Assignment Sheet: Energy Descent Action Plan
[used for students doing the Applied Permaculture module at Kinsale College, Ireland]

This is your major piece of coursework for the Applied Permaculture module and it comprises 50% of your overall mark for the module. You are to contribute to two chapters of the finished Action Plan report. This project is your opportunity to pull together everything you have learned over your years as a permaculture student. Go for it: this is a piece of work that people will talk about in the future as being where it all started in Kinsale!

Each chapter should follow the following format:

Title (e.g.) Food, Tourism, whatever.

Authors Your names

The Present This section should be no longer than a single paragraph (six to eight sentences) and should sum up the current situation and its problems, as well as the issues posed by peak oil with regard to the particular topic.

The Vision This section should not exceed two paragraphs in length, and should look at how, in your vision, Kinsale will be in relation to your subject area by 2021. Envisage Kinsale with only 25% of the fossil fuels currently available. How do you see Kinsale, in that scenario, having adapted to it in such a way as to actually be a more pleasant place to live? What would it look like? What would it be like to live there? This section needs to communicate clearly and passionately your vision for a post-carbon Kinsale.

Practical Steps This is the main part of this project. In this section you are to set out, in a chronological order, the practical steps that need to be undertaken to move towards your vision. How, starting from today, will you begin to make the move towards the sustainable low-energy Kinsale, in such a way that most of the community are on board with the process? How can you pull in and integrate the existing organisations in Kinsale? Who will undertake the work you are proposing? If you are able to state whether a particular step should be carried out by the community, the Town Council or the County Council, then put that in. Your proposals need to be bold but also achievable. Draw on best practice where you can find good examples, you can cite the sources for these in your final section (see below). Your steps should each be dated.

Resources Here you can list any useful websites you used for research, any books, and any useful organisations. The idea with this section is that anyone reading the report can follow up areas in which they have an interest. You can put international links, but also links and contact details for organisations already working in the area (however ineffectual you may feel they are), such as Sustainable Energy Ireland or the Waste Management people at Cork County Council.

Also If you have any photographs or illustrations you would like to include, then do. I have a number of pictures of the course and of trips we have been on, which can be included in the final report, but any other suggestions for graphics would be very much welcomed.

The essence of this report is that it is concise, readable, and that it stimulates the readers with loads of ideas that they previously would never have considered. Your task here is to distil all that you know about permaculture and sustainable solutions and apply it to a real-life settlement. This is town-scale permaculture. Make it bold and exciting so that it can seize people's imaginations.

Hand-in Date You are to present a draft version of your chapters by April 13th 2005. On that day we will undertake a group editing process, and all read each others chapters and make comments. These will then be amended, also with comments from myself, and ready for final submission on April 27th 2005. Your submission MUST be word-processed and either emailed to me, or given to me on a floppy or a CD.

I will then edit and set the whole document, which we will get printed before the end of term.

Appendix 4

The Energy Descent Action Plan process

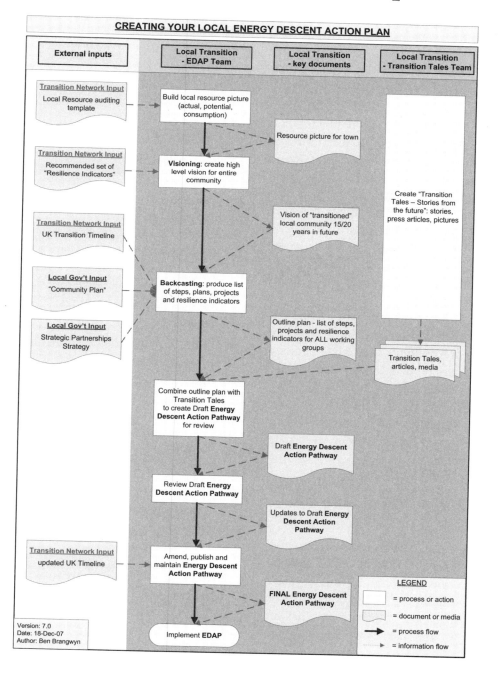

CREATING YOUR LOCAL ENERGY DESCENT ACTION PLAN

External inputs	Local Transition - EDAP Team	Local Transition - key documents	Local Transition - Transition Tales Team

Transition Network Input
Local Resource auditing template

Build local resource picture (actual, potential, consumption)

Resource picture for town

Transition Network Input
Recommended set of "Resilience Indicators"

Visioning: create high level vision for entire community

Vision of "transitioned" local community 15/20 years in future

Create "Transition Tales – Stories from the future": stories, press articles, pictures

Transition Network Input
UK Transition Timeline

Local Gov't Input
"Community Plan"

Backcasting: produce list of steps, plans, projects and resilience indicators

Outline plan - list of steps, projects and resilience indicators for ALL working groups

Local Gov't Input
Strategic Partnerships Strategy

Transition Tales, articles, media

Combine outline plan with Transition Tales to create Draft **Energy Descent Action Pathway** for review

Draft **Energy Descent Action Pathway**

Review Draft **Energy Descent Action Pathway**

Updates to Draft **Energy Descent Action Pathway**

Transition Network Input
updated UK Timeline

Amend, publish and maintain **Energy Descent Action Pathway**

FINAL **Energy Descent Action Pathway**

Implement **EDAP**

Version: 7.0
Date: 18-Dec-07
Author: Ben Brangwyn

LEGEND
☐ = process or action
⬭ = document or media
➡ = process flow
⇢ = information flow

Appendix 5

How to become a Transition Initiative

by Ben Brangwyn, co-ordinator of the Transition Network

Having read this book, you may feel inspired either to initiate your own Transition Initiative or to get involved with an existing one. If so, take some comfort that you will not be the first place to do it. Already there is a rapidly evolving network of such Initiatives around the UK and beyond. The Transition Network has been established to inspire, inform, support, network and train communities as they consider, adopt and implement a Transition Initiative. It is currently building a range of materials, training courses, events, tools and techniques, resources and a general support capability to help these communities.

As a critical mass of communities embarking on this energy descent planning process emerges, we are building a thriving co-operative network where people are sharing best practice, helping each other and creating a way of life that is far better than the atomised, disconnected unsustainable and inequitable society that we've grown into, largely on the back of super-abundant cheap oil.

Setting up your Transition Initiative

We've established a draft set of criteria that tells us how ready a community is to embark on this journey to a lower-energy future. If you're thinking of adopting the Transition model for your community, take a look at this list and make an honest appraisal of where you are on these points. If there are any gaps, it should give you something to focus on while you build the initial energy and contacts around this Initiative.

We've introduced this somewhat formal approach to registering Transition Initiatives for several key reasons:

- Our trustees and funders want to make sure that while we actively nurture embryonic projects, we only promote to 'official' status those communities we feel are ready to move into the awareness-raising stage. This status confers additional levels of support such as speakers, trainings, Wiki and forums that we're currently rolling out

- In order to establish co-ordinated programmes (such as combined funding bids to the National Lottery) we need a formally established category of Transition Initiatives that we're fully confident can support and deliver against such programmes

- We've seen at least one community stall because they didn't have the right mindset or a suitable group of people, and didn't really understand what they were letting themselves in for.

The communities that have gone through this initial formal process have universally agreed that it has enhanced their capacity to build a robust and well-supported initiative.

Criteria for becoming a Transition Initiative

These criteria are developing all the time, and certainly aren't written in stone, but they are designed to be as helpful to you as possible.

- An understanding of Peak Oil and Climate Change as twin drivers (to be written into your group's constitution or governing documents)

- A group of four or five people willing to step initially into central roles (not just the boundless enthusiasm of a single person)

- At least two people from the core team willing to attend an initial two-day training course. These courses will be held both in Totnes and regionally

- A potentially strong connection to the local council

- An initial understanding of the Twelve Steps of Transition

- A commitment to ask for help when needed

- A commitment to regularly update your Transition Initiative web presence – either the Wiki (collaborative workspace on the web that we'll make available to you), or your own website

- A commitment to write up something on the Transition Network blog once every couple of months (the world will be watching . . .)

- A commitment, once you're into the Transition, for your group to give at least two presentations to other communities (in the vicinity) that are considering embarking on this journey – a sort of 'here's what we did' or 'here's how it was for us' talk

- A commitment to network with other communities in Transition

- Minimal conflicts of interests in the core team

- A commitment to work with the Transition Network re grant applications for funding from national grant-giving bodies. Your own local trusts are yours to deal with as appropriate

- A commitment to strive for inclusivity across your entire Initiative

- A recognition that although your entire county or district may need to go through Transition, the first place for you to start is in your local community. It may be that eventually the number of Transitioning communities in your area warrant some central group to help provide local support, but this will emerge over time, rather than be imposed. This point is in response to the several instances of people rushing off to Transition their entire county/region rather than their local community.

In exceptional situations where a co-ordinating hub or initiating hub needs to be set up (currently Bristol, Forest of Dean, Brighton & Hove, Nottingham) that hub is responsible for making sure these criteria are applied to all the Initiatives that start within their area. Further responsibilities for ongoing support and possibly training are emerging as we see this role develop.

Further criteria apply to initiating/co-ordinating hubs – these can be discussed person to person.

- Finally, we recommend that at least one person on the core team should have attended a permaculture design course. It really does seem to make a difference.

Once you can let us know that you're on board with these and ready to set off on your Transition journey, you open the door to all sorts of wonderful support, guidance, materials, web space, training, networking opportunities and co-ordinated funding initiatives.

Setting up your Transition Initiative: the different types

The three types of Initiatives we have identified thus far are:

- The 'Local Transition Initiative', embedded in its own locale, where the steering group inspires and organises the local community. This is the most frequent and easy to handle initiative, typically with communities of up to 15,000 people. Examples of this include Totnes, Lewes, Wrington.

- The 'Local Transition Hub' based within a large congruent/contiguous area with its own identity (e.g. a city). That group's role is to fire up Transition Initiatives in the surrounding area and maintain a role of inspiring, encouraging, registering, supporting, networking and possibly training those Initiatives. Examples of this include Bristol, Brighton & Hove and Forest of Dean. The Local Transition Hub will handle pretty well all the communications with the local Initiatives in that area.

- the 'Temporary Initiating Hub' made up of a collection of individuals/groups from separate locales in the same region who are accustomed to working with each other in some kind of activist/environmental

capacity. This group helps each other to fire up Local Transition Initiatives in the region and then dissolves, with the original members moving into their own Local Transition Initiatives once they've achieved some critical mass to form a local steering group. West Berkshire and a couple of others are taking this approach. In this scenario, the local Initiatives, once they've started up, will look to the Transition Network for support, training etc.

More about the 'Local Transition Hub' role

If a group wants to take the role of a Local Transition Hub, then we at the Transition Network need to be really confident that you know what you're letting yourself in for and that you're going to be able to handle it. This'll probably involve a number of conversations, and probably a face-to-face discussion with the team.

We're planning to set up a 'Local Transition Hub' community to discuss the complexities of this approach – and there are many. I think this Local Transition Hub model is crucial for the cities, but the early adopters are going to have to feel their way carefully into this role. It's virgin territory and by no means a trivial undertaking, so proceed with caution.

We look forward to hearing from you and to working with you on this exciting process. To find out more, contact the Transition Network at:

43 Fore Street, Totnes, Devon TQ9 5HN
Tel: 0560 1531882
benbrangwyn@transitionnetwork.org

Appendix 6

A sample press release

Press Release – For Immediate Release

From Transition Town Totnes, 43 Fore Street, Totnes, Devon TQ9 5HN. 31st August 2007

Transition Town Totnes Announces New Programme of Events

Transition Town Totnes, the project exploring the town's transition to life after oil has just unveiled its new Programme of Events for the next four months. It all kicks off on Thursday 6th September with TTT's First Birthday Party at the Seven Stars Hotel, Totnes, celebrating the wonderful successes and achievements of the first TTT year. An evening of music, reflection, comedy, conversation, poetry, and dancing, it features poet Matt Harvey, Bert Miller and the Animal Folk, Rooh Star, clowns, Breton Dance Band Poisson Rouge, canny competitions, a ravishing raffle, and many more surprises.

TTT's founder Rob Hopkins told the *Totnes Times*, "Season Two's programme promises a dazzling array of evening events and Open Space days, looking at the potential futures of transport, education, and agriculture beyond the availability of cheap oil and gas, as well as a day specifically for young people. We think this is our strongest programme yet and that it is a testament to the energy and dedication the community has put behind this initiative."

Other highlights include talks by Herbert Girardet, David Strahan, author of *The Last Oil Shock*, Paul Allen, the director of the Centre for Alternative Technology in Wales, Bernard Lietaer, Richard Douthwaite, Peter Lipman of Sustrans and Craig Sams, founder of Green and Black's chocolate, along with many more.

There will also be a performance of 'This Farming Life', an extraordinary combination of traditional song, storytelling, and archive footage of farming in the area as far back as the 1920s, and an Interfaith event at St Johns Church, as well a programme of Great Reskilling courses and workshops offering the possibility of learning new practical skills. Are you longing to make your own YouTube video? Passionate to preserve your harvest or to learn the fine art of bread-making? From rocket-stove making to learning about local clays and seed saving. . . the new programme of events has everything you could wish for this autumn.

In a birthday greeting to TTT, Andrew Simms, Director of the New Economics Foundation, who spoke in Totnes recently, wrote:

"Whether we like it or not the world faces a major transition away from dependence on fossil fuels. Either we will decide to manage that transition for the maximum benefit of everyone, or we will be managed by it in a way that is likely to be messy, dangerous and potentially lethal for millions. By taking this initiative, Totnes is leading by example. Transition Town Totnes is a pioneer. It is one of the first beats in a drum roll announcing the arrival of a new era in which we will come to understand that good lives do not have to cost the earth."

The Season Two Programme of Events is widely distributed throughout Totnes, and on information boards in the Library, Willow Cafe, Greenlife Shop, and of course at the new TTT Office, 43 Fore Street, Totnes, which is open 10am-3pm Tuesday-Friday, from September 10th.

The One Year TTT Anniversary Birthday Party will be at the Seven Stars Hotel, Totnes, 8-11.30pm. Advance tickets are available from Greenfibres and Totnes Pet and Garden, £6 (£5 concessions). Raffle tickets can be bought at the Red Wizard Cafe, Riverford Goes To Town, and the Harlequin Bookshop. Totnes pounds will be accepted for all TTT events.

For more information about these events please contact the TTT Office on 01803 867358 or visit the TTT website at www.transitiontowns.org/Totnes. Email: transitiontowntotnes@googlemail.com

Transition Training

Transition Training has emerged to meet the growing need from communities for training in the thinking and practical tools required by emergent Transition Initiatives. We offer a range of training courses, materials and events to support and develop the work of Transition Initiatives everywhere.

Set up in 2007, our initial projects have been to offer some basic training in the Transition model and process, and to research training needs across the Network. During 2008 we will develop a wider programme, involve more people in forming and delivering the training, and support Transition projects in setting up their own training initiatives. We welcome input and suggestions for this process.

Two courses are already available:

Training for Transition

This workshop is designed to give a detailed introduction to the most important skills necessary to successfully set up, develop, and run a Transition project in your locality. Aimed at those who are considering or already initiating a local Transition project, it covers all the following areas:

- the context for Transition projects, the current global situation and the transformational possibilities that arise from climate change and peak oil and gas

- the principles underlying the Transition model

- the process for starting a successful Transition project

- how a Transition project could move forward in your area

The workshop is participatory, action-learning-based, and fun, with participants invited to share their own experience and learn from the many different Transition projects represented in the group.

How to give a Transition Town Talk

An important part of the Transition process is to inform people about what is happening in the world and inspire them with a vision for a positive future. We have distilled the expertise of those who have given such talks to businesses, communities, conferences, government groups and even royalty and created a one-day training in how to give an informative, entertaining and motivating talk.

The initiators of Transition Training are **Naresh Giangrande** *and* **Sophy Banks**. *Focaliser of the Energy Group and one of the founders of the Transition Town Totnes project, Naresh has been involved in designing, running and evolving many of the events and groups that have been at the heart of the TTT Initiative. Sophy jointly focalises the 'Heart and Soul/psychology of change' group as well as co-ordinating the central office and systems. A trainer for over twenty years, she has worked as an engineer, information systems consultant and psychotherapist and has considerable experience of voluntary sector and community projects.*

New courses are being developed all the time. More information can be found at www.transitiontowns.org/TransitionNetwork/TransitionTraining.

For more information call 05601 531882 or email training@transitionnetwork.org.

References

Part One & Chapter 1: Peak Oil and Climate Change

1. Heinberg, R. (2006) *The Oil Depletion Protocol: A Plan to Avert Oil Wars, Terrorism and Economic Collapse*, Clairview Publications.

2. Chris Skrebowski is the editor of the *Petroleum Review*. This quote comes from a conversation between him and Richard Heinberg, April 2006.

3. A term originally coined by Jeremy Leggett.

4. A selection of recommended reading on peak oil appears in the Resources section (p. 233). An excellent and very concise introduction is Heinberg, R. (2007) *The View from Oil's Peak*. www.richardheinberg.com/museletter/184.

5. Dr Seuss (2004) *The Lorax*, Collins.

6. For a concise overview of how oil was formed, see pages 29-51 of *Oil Crisis* by Dr Colin Campbell (2005), Multi-Science Publishing Ltd.

7. Siegel, L. (2003) *Bad Mileage: 98 tons of plants per gallon: study shows vast amounts of 'buried sunshine' needed to fuel society*, University of Utah. www.eurekalert.org/pub_releases/2003-10/uou-bm9102603.php.

8. FEASTA (2007) *The Great Emissions Rights Giveaway*. www.feasta.org.

9. Ibid.

10. Frances, H. (2007) from a presentation entitled 'Global Energy Demand Trends' given at the ASPO 6 conference, Cork, Ireland, 17th September 2007.

11. Heinberg, R. (2003) *The Party's Over: oil, war and the fate of industrial societies*, Clairview Books, p.31.

12. Ibid.

13. Ibid, p.44.

14. From a presentation at Dartington Hall at an event called 'Food and Farming in Transition', organised by Transition Town Totnes. You can see his actual presentation at www.transitionculture.org/2007/06/12/chris-skrebowski-at-estates-in-transition/.

15. Hubbert's story is told in more detail in Strahan, D. (2007) *The Last Oil Shock: The Imminent Extinction of Petroleum Man*, John Murray Publishing.

16. Cohen, D. (2007) *The Perfect Storm*. ASPO-USA/Energy Bulletin. 31st October 2007. www.energybulletin.net/36510.html.

17. Energy Watch Group (2007) *Crude Oil: The Supply Outlook*. EWG Series No.3. On page 46 is the following section: "The growth of production has come to a standstill and production now is more or less on a plateau. This has happened despite historically high oil prices."

18. Strahan, D. (2007) *The Last Oil Shock: The Imminent Extinction of Petroleum Man*, John Murray Publishing, p.62.

19. Energy Watch Group (2007) *Crude Oil: The Supply Outlook*. EWG Series No.3, p.101.

20. This figure is quite variable, with estimates ranging from three barrels (Strahan, D., 'What Stern really got wrong', *Prospect*, 16th May 2007), to others who say it is six barrels or more to every one discovered. In an email exchange, Colin Campbell of ASPO states that total discoveries in 2006 amounted to 5.2 Gb and world consumption according to BP was 30.56 Gb, resulting in a ratio of 1:6, although he states that uncertainty about the exact discovery figures add a degree of uncertainty (personal communication by email, 13th September 2007).

21. More information about the Hedberg Conference can be found in Strahan, D. (2007) *Private Industry Conference finds much less oil*. www.LastOilShock.com. Friday 28th September 2007.

22. Mawdsley, J., J. Mikhareua and J. Tennison (2001) *The Oil Sands of Canada: the world wakes up: first to peak oil, second to the oil sands of Canada*, Equity Research of Canada, Raymond James.

23. Taken from a piece I wrote reviewing an episode of BBC Radio 4's 'In Business' programme which explored the Alberta Tar Sands. ww.transitionculture.org/2006/01/27/the-alberta-oil-rush-on-radio-4/.

24. Greenpeace Canada (2007) *Questions and Answers about the Alberta Tar Sands*. www.greenpeace.org/canada/en/recent/tarsandsfaq#4.

25. See for example Darley, J. (2004) *High Noon for Natural Gas: The New Energy Crisis*, Chelsea Green Publishing Company.

26. Simmons, M. (2006) *Tough Times Ahead for Energy*. Financial Sense Newsletter. www.financialsense.com/transcriptions/2006/0429Simmons.html.

27. Nelder, C. (2007) *Tar Sands: The Oil Junkie's Last Fix, Part Two*, The Oil Drum, 9th September 2007, www.canada.theoildrum.com/node/2931. The article is a devastating critique of tar sands production.

28. For example, Morrison, M. (2007) *Plenty of Oil — Just Drill Deeper: The discovery of reserves in the Gulf of Mexico means supply isn't topping out*, BusinessWeek.com, 7th September 2006.

29. Cohen, D. (2006) *Jack-2 and the Lower Tertiary of the Deepwater Gulf of Mexico*. Posted at The Oil Drum.com. www.theoildrum.com/story/2006/9/8/11274/83638.

30. International Energy Agency (2007) *Medium Term Oil Market Report*. 10th August 2007. www.omrpublic.iea.org/currentissues/full.pdf. A very useful analysis of this report is Cohen, D. (2007) *Inside the IEA's Medium Term Oil Market Report*, ASPO USA. www.aspo-usa.com.

31. Leonard, A. (2007) *If It Smells Like Peak Oil, It Probably Is*. www.salon.com/tech/htww/2007/07/09/iea_report/index.html.

32. Strahan, D. (2007) 'Why BP and Shell are bound to merge', *The Independent*, Sunday 15th July 2007.

33. Pauly, D. (2007) *Slow, Steady Liquidation of the World Oil Industry*, Bloomberg.com. 1st October 2007. www.bloomberg.com/apps/news?pid=20601039&sid=akIQ2arQB4Qs&

refer=columnist_pauly.

34. John S. Herold Inc. (2007) *2007 Global Upstream Performance Review*, John S. Herold Inc. & Harrison Lovegrove & Co.

35. A day rarely passes without my regularly checking www.energybulletin.net and www.theoildrum.com.

36. The seminal reference on this is Simmons, M. (2005) *Twilight in the Desert: The Coming Saudi Oil Shock and the World Economy*, John Wiley and Sons. Also useful is Stuart Staniford's ongoing analysis of Saudi production at www.theoildrum.com.

37. Monbiot, G. (2005) 'Crying Sheep: we had better start preparing for a decline in global oil supply'. *The Guardian*, 27th September 2005.

38. Homer-Dixon, T. (2007) *The Upside of Down: Catastrophe, Creativity and the Renewal of Civilisation*, Souvenir Press.

39. Deffeyes, K. (2006) *Beyond Oil: The View from Hubbert's Peak*, Hill & Wang.

40. You can hear the audio of his talk, entitled *The Peak of World Oil Production: Thanksgiving Day, 2005* at www.globalpublicmedia.com. He also made this prediction in a piece called *Join Us As We Watch The Crisis Unfolding*, posted at www.princeton.edu/hubbert/current-events-05-11.html.

41. Heinberg, R. (2007) *The View from Oil's Peak*. Museletter #184, August 2007. www.richardheinberg.com/museletter/184.

42. Oil Depletion Analysis Centre (2004) *Oil field mega projects: E&P Review*. Available from www.odac-info.org/bulletin/documents/MEGAPROJECTSREPORT.pdf.

43. Campbell, C. (2002) *Peak Oil: an Outlook on Crude Oil Depletion*. www.greatchange.org/ov-campbell,outlook.html.

44. Chris Skrebowski, Editor of *Petroleum Review*, interviewed by Julian Darley, 11 April 2005. www.globalpublicmedia.com/interviews/378.

45. Laherrère, Jean. (2003) 'Hydrocarbons Resources Forecast of oil and gas supply to 2050', *Petrotech Conference*, New Delhi. www.hubbertpeak.com/laherrere/Petrotech090103.pdf.

46. Jackson, P.M. (2007) *Why the 'Peak Oil' Theory Falls Down: Myths, Legends and the Future of Oil Resources*. Cambridge Energy Research Associates.

47. Cohen, D. (2007) *Does the Peak Oil 'Myth' Just Fall Down?: Our Response to CERA*. www.theoildrum.com/story/2006/11/15/83857/186.

48. Mortished, C. (2006) 'Total chief says world will find oil target tough', *The Times*, 8th September 2006.

49. Bergin, T. (2006) 'Total sees peak oil output around 2020', Reuters. Archived at www.energybulletin.net/16850.html.

50. Strahan, D. and Watson. A. (2007) 'Oil industry sleep walking into crisis', *The Independent*, 17th September, 2007.

51. Energy Watch Group (2007) *Crude Oil; The Supply Outlook. Report to the Energy Watch Group*. October 2007. EWG-Series No 3/2007.

52. Leake, J. (2007) 'The Road Fix', *Sunday Times*, 9th August, 2007.

53. Prime Minster's Office (2007) 'The Government's response to the Peak Oil Petition'. 3rd October 2007, www.pm.gov.uk/output/Page13388.asp.

54. www.globalcool.org.

55. Astyk, S. (2007) 'How fast is global warming happening?' www.casaubonsbook.blogspot.com, 27th September, 2007.

56. McCarthy, M. (2007) 'What's happening to our weather?', *The Independent*, Tuesday 28th August, 2007.

57. Leake, J. & Carpenter, J. (2007) 'Britain's Atlantis Under the North Sea', *Sunday Times*, 2nd September 2007.

58. Hansen, J., L. Nazarenko, R. Ruedy, Mki. Sato, J. Willis, A. Del Genio, D. Koch, A. Lacis, K. Lo, S. Menon, T. Novakov, Ju. Perlwitz, G. Russell, G. A. Schmidt, and N. Tausnev. (2005) 'Earth's Energy Imbalance: Confirmation and implications', *Science*, 308, pp.1431-1435.

59. Brahic, C. (2007) 'Sea level rise outpacing key predictions', *New Scientist* News Service, 1st February 2007. www.environment.newscientist.com/article/dn11083.

60. The Intergovernmental Panel on Climate Change's Four Assessment Report, published in 2007, states that it has "very high confidence" (defined as being at least a 9 out of 10 chance of being correct) that "the global net effect of human activities since 1750 has been one of warming." IPCC Summary for Policymakers, p.3.

61. Lynas, M. (2007) *Six Degrees: our future on a hotter planet*, Fourth Estate.

62. McKie, R. (2007) 'Arctic thaw opens fabled trade route', *The Observer*, 16th September 2007.

63. Jha, A. (2006) 'Boiled Alive'. *The Guardian*, 26th July 2006.

64. Monbiot, G. (2006) *Heat: how to stop the planet burning*, Penguin.

65. Hansen, J, Sato, M, Kharecha, P, Russell, G, Lea, D. W. & Siddall, M. (2007) 'Climate change and trace gases', *Philosophical Transactions of The Royal Society A*, Volume 365, Number 1856, 15th July 2007. www.journals.royalsoc.ac.uk/content/l3h462k7p4068780/fulltext.html.

66. IPCC (2007) *Global climate projections, Climate Change 2007: The Physical Sciences Basis*. www.ipccwg1.ucar.edu/wg1/wg1-report.html.

67. Spratt, D. (2007) 'The Big Melt: lessons from the Arctic Summer of 2007', *Carbon Equity*. Available from www.carbonequity.info. One of the most chilling things you will ever read.

68. Bjornes, C. (2007) *International polar day*, Cicero, 17th September 2007. www.cicero.uio.no/webnews/index_e.aspx?id=10868.

69. Revkin, A. C. (2007) 'Retreating Ice: A blue Arctic Ocean in summers by 2013?' *International Herald Tribune*. 1st October 2007. www.iht.com/articles/2007/10/01/sports/arcticweb.php.

70. Kahn, M. (2007) 'Sudden sea level surges threaten 1 billion', Reuters, 19th April 2007. www.uk.reuters.com/article/scienceNews/idUKN1941671620070419.

71. Hansen, J., L. Nazarenko, R. Ruedy, Mki. Sato, J. Willis, A. Del Genio, D. Koch, A. Lacis, K. Lo, S. Menon, T. Novakov, Ju. Perlwitz, G. Russell, G.A. Schmidt, and N. Tausnev. (2005) 'Earth's energy imbalance: Confirmation and implications', *Science*, 308, 1431-1435.

72. Spratt, D. (2007) 'The Big Melt: lessons from the Arctic Summer of 2007', *Carbon Equity*. Available from www.carbonequity.info., p.13.

73. The figure around which George Monbiot's book *Heat: how to stop the planet burning* is built.

74. Global Commons Institute (2007) 'A Briefing for Channel Four'. www.gci.org.uk/briefings/Channel_Four.pdf.

75. One of the best, most insightful studies on where the safe limit lies, which came out just as this book was in its final editing, is Spratt, D. & Sutton, P. (2007) 'Target Practice: where should we aim to avoid dangerous climate change?', Carbon Equity/ GreenLEAP Strategic Institute. Available from www.carbonequity.info/PDFs/targets.pdf.

76. From a transcript of a response he gave at an event in Lampeter. Posted on Transition Culture on April 10th 2007, entitled *George Monbiot on Peak Oil and Transition Towns*. www.transitionculture.org/2007/04/10/george-monbiot-on-peak-oil-and-transition-towns.

77. Ibid.

78. Monbiot, G. (2007) 'What if the oil runs out?', *The Guardian*, 29th May 2007. www.monbiot.com.

79. Interview with Tony Juniper, www.transitionculture.org, 23rd February 2007.

80. Leggett, J. (2006) 'Peak time viewing', *The Guardian*, 15th March 2006.

81. Environmental News Service (2007) *New York Tallies Its Greenhouse Gas Emissions*. www.ens-newswire.com/ens/apr2007/2007-04-11-03.asp.

82. BBC News Online (2006) 'Sweden aims for oil-free economy'. Wednesday, 8 February 2006. www.news.bbc.co.uk/1/hi/sci/tech/4694152.stm.

83. Hopkins, R. (2007) 'In Praise of ASPO 6. #4. Eamon Ryan'. www.transitionculture.org/2007/09/26/aspo-6-in-praise-of-5-eamon-ryan/.

84. Heinberg, R. (2006) *The Oil Depletion Protocol: A Plan to Avert Oil Wars, Terrorism and Economic Collapse*, Clairview Books.

85. Stern, N., S. Peters, V. Bakhshi, A. Bowen, C. Cameron, S. Catovsky, D. Crane, S. Cruickshank, S. Dietz, N. Edmonson, S.-L. Garbett, L. Hamid, G. Hoffman, D. Ingram, B. Jones, N. Patmore, H. Radcliffe, R. Sathiyarajah, M. Stock, C. Taylor, T. Vernon, H. Wanjie, and D. Zenghelis (2006) *Stern Review: The Economics of Climate Change. Part III The Economics of Stabilisation*. HM Treasury, London, p.185.

86. Heinberg, R. (2007) 'Bridging Peak Oil and Climate Change Activism', Museletter # 177.

January 2007. www.richardheinberg.com/museletter/177.

87. Nelder, C. (2007) 'Peak Oil Hits the Third World: high oil prices bring energy shortages'. Energyandcapital.com, 10th August 2007. www.energyandcapital.com/articles/peak+oil-renewable+energy-shortages/490.

88. Hirsch, R.L., Bezdek, R. Wendling, R. (2005a) 'Peaking of World Oil Production: Impacts, Mitigation and Risk Management', National Energy Technology Laboratory, US Department of Energy.

89. Hirsch, R. L. (2005) 'Robert Hirsch on Peak Oil Mitigation', Global Public Media. www.globalpublicmedia.com/transcripts/2459.

90. Interview with Richard Heinberg in Stroud, Gloucestershire, March 2007.

91. Hopkins, R. (2006) 'ASPO 5: Robert Hirsch Scares Me Out of My Wits', Transition Culture, August 18th 2006.

92. Brown, L. (2003) *Plan B: Rescuing a Planet Under Stress and a Civilization in Trouble*, W. W. Norton/Earth Policy Institute.

Chapter 2: The view from the mountain-top

1. Holmgren's four scenarios are set out in Holmgren, D. (2005) 'The End of Suburbia or the Beginning of Mainstream Permaculture?', *Permaculture Magazine* 46, pp.7-9.

2. Kleiner, A. (1996) *The Age of Heretics*, Nicholas Brealey Publishers.

3. Gallopin, C.G. (2002) 'Planning For Resilience: Scenarios, Surprises and Branch Points', in Gunderson L.H. and C.S. Holling, *Panarchy: Understanding Transformations in Human and Natural Systems*, Washington, Island Press; Holmgren, D. (2003) 'What is Sustainability?', Sustainability Network Update 31E, 9th September 2003; Heinberg, R. (2004), *Powerdown: Options and actions for a Post-Carbon World*, Clairview Books; Curry, A., T. Hodgson, R., Kelnar and A. Wilson (2005) *Intelligent Future Infrastructure: The Scenarios Towards 2055*, Foresight, Office of Science of Technology, FEASTA (2006) *Energy Scenarios Ireland*. FEASTA.

4. For a concise summary of this argument see, for example, Howden, D. (2007) 'World oil supplies are set to run out faster than expected, warn scientists', *The Independent*, 14th June 2007 or Heinberg, R. (2007) '*The View from Oil's Peak*', Museletter #184, August 2007. www.richardheinberg.com/museletter/184.

5. Diamond, J. (2006) *Collapse: How Societies Choose to Fail or Survive*, Penguin.

6. Catton, W.R. (1982) *Overshoot: The Ecological Basis of Revolutionary Change*, University of Illinois Press.

7. The term 'Lean Economy' was developed by David Fleming in 1995, and he has published on the concept regularly since then, e.g. *After Growth – Climax: Rising Unemployment as the Cue for Evolution to the Lean Economy*, European Environment (1998) vol 8, pp.41-49. The idea was originally derived from the Japanese system of 'lean production'. His book, *Lean Logic: The Book of Environmental Manners* is forthcoming. See www.theleaneconomyconnection.net.

8. The term 'Powerdown' comes from Heinberg, R. (2004) *Powerdown: options and actions for a post-carbon world*, Clairview Books.

9. City of Portland Peak Oil Task Force (2007) 'Descending the Oil Peak: Navigating the Transition from Oil and Natural Gas'. Download from www.portlandonline.com/shared/cfm/image.cfm?id=145732.

10. Shiva, V. (2005) 'Tsunami Teachings: Reflections for the New Year'. www.zmag.org, 23rd January 2005.

11. Winter, D. D. and S.M. Kroger (2004) *The Psychology of Environmental Problems*, New Jersey, Lawrence Erlbam Associates.

12. Korten, D.C. (2000) *The Post Corporate World: Life After Capitalism*, Berrett-Koehler Publishers.

13. Abdullah, S. (1999) *Creating a World That Works For All*, Berrett-Koehler Publishers.

14. I don't remember where I first read this quote, and rooting around on the internet failed to provide a source. It can however be found, along with other quotes by Debuffet, at www.quote.robertgenn.com/auth_search.php?authid=636.

15. FEASTA (2006) 'The Great Emissions Rights Give Away', FEASTA, March 2006.

16. Hall, C. (2008) 'Provisional Results from EROI Assessments', State University of New York (draft).

17. Palcer, S., M. C. Herweyer and C. Hall (2008) 'Crude Oil Imported to the United States'. Appendix to the above.

18. Nelder, C. (2007) 'Peak Oil Hits the Third World: High Oil Prices Bring Energy Shortages', www.Energyand Capital.com, 10th August 2007. www.energyandcapital.com/articles/peak+oil-renewable+energy-shortages/490.

19. Explored in more depth in Heinberg, R. (2007) *Peak Everything: Waking Up to the Century of Declines*, New Society Publishers.

20. Although Charles Hall's 2008 paper looks likely to be the most thorough yet produced, important previous papers on the subject include Cleveland, C. J., R. Constanza, C. A. S. Hall, and R. Kaufmann (1984) 'Energy and the U.S. Economy: A Biophysical Perspective', *Science* 225: 890-897; Hall, C. A. S., C. J. Cleveland, and R. Kaufmann (1986) *Energy and Resource Quality: The Ecology of the Economic Process*, John Wiley & Sons, Inc. pp.221-228; and Cleveland, C. (2005) 'Net energy obtained from extracting oil and gas in the United States', *Energy* 30:769-782.

21. Odum, H.T. & Odum, E.C. (2001) *A Prosperous Way Down: principles and policies*, University Press of Colorado.

22. Hirsch, R.L., Bezdek, R., Wendling, R. (2005a) *Peaking of World Oil Production: Impacts, Mitigation and Risk Management*, National Energy Technology Laboratory, US Department of Energy.

23. Holmgren, D. (2003b) *What Is Sustainability?* from Sustainability Network Update 31E, 9th September 2003, CSIRO Sustainability Network. www.bml.csiro.au/susnetnl/netwl31Eb.pdf.

24. Trainer, T. (2007) *Renewable Energy Cannot Sustain a Consumer Society*, Springer Verlag.

25. Ibid. p.9.

26. First appeared in Hopkins, R. (2007) 'Energy Descent Pathways: evaluating potential responses to peak oil'. An MSc dissertation for the University of Plymouth. Available from www.transitionculture.org.

Chapter 3: Why rebuilding resilience is as important as cutting carbon emissions

1. Walker, B, Hollinger, C.S., Carpenter, S.R. and Kinzig, A. (2004) 'Resilience, Adaptability and Transformability in Social-ecological Systems', *Ecology and Society* 9 (2) p.5.

2. For a more detailed overview of the lorry drivers' dispute, see Strahan, D. (2007) *The Last Oil Shock: A Survival Guide to the Imminent Extinction of Petroleum Man*, John Murray.

3. Department of Environment, Food and Rural Affairs. Press statement released at the 2003 Royal Show.

4. Fleming, D. (2007) *Lean Logic, a Dictionary of Environmental Manners*. Unpublished MS.

5. For example Levin, S. A. (1999) *Fragile Dominion*, Perseus Books Group.

6. For a thorough demolition of the concept of monoculture, see Shiva, V. (1998) *Monocultures of the Mind: Biodiversity, Biotechnology and Scientific Agriculture*, Zed Books.

7. Thanks to David Fleming for identifying these two points in an email exchange.

8. Walker, B. & Salt, D. (2007) *Resilience Thinking: Sustaining Ecosystems and People in a Changing World*, Island Press.

9. Ibid. p.121.

10. New Economics Foundation (2007) 'Clone Town Britain: The survey results on the bland state of the nation'. www.neweconomics.org/gen/z_sys_publicationdetail.aspx?pid=206.

11. Leopold, A. (1972) *Round River*, Oxford University Press.

12. Kunstler, J. J. (2005) *The Long Emergency: surviving the converging catastrophes of the twenty-first century*, Atlantic Monthly Press.

13. For example, see the DVD *The End of Suburbia: Oil Depletion and the Collapse of the American Dream*, Electric Wallpaper Company.

14. Many thanks to George Heath's son, David, for his time in exploring the history of Heath's Nursery, and also to Alan Langmaid of Totnes Museum and to the Totnes Image Bank and Rural Archive for their insights.

15. Dartington Rural Archive (1985) *Horsepower*, Spindle Press.

16. For a detailed history of coal, see Freese, B. (2006) *Coal: A Human History*, Arrow Books.

17. I read this quote somewhere years ago, and wrote it in a notebook attributing it to Richard Mabey. When putting this book together I decided I wanted to use it, and emailed Mabey to ask for the source of the quote, only for him to write back saying "That sentence ain't by me!" So, I have no idea who said it – perhaps I made it up. If anyone can enlighten me as to who said this, I'll be very grateful, as it is one of my favourite quotes.

18. Simms, A. (2005) *Ecological Debt: The Health of the Planet and the Wealth of Nations*, Pluto Press.

19. Wilt, A.F. (2001) *Food for War: Agriculture and Rearmament in Britain before the Second World War*, Oxford University Press.

20. Gardiner, J. (2004) *Wartime Britain 1939-1945*, Headline Book Publishing.

21. Ibid.

22. Hammond, R.J. (1954) *Food and Agriculture in Britain 1939-45*, Food Research Institute. Stanford University Press. A hard-to-track down title, but a fascinating and authoritative account of the UK Government's preparations for war.

Chapter 4: Why small is inevitable

1. Fleming, D. (2007) *Lean Logic: The Dictionary of Environmental Manners*. Forthcoming.

2. Talberth, J. et.al. (2006) *Building a Resilient and Equitable Bay Area: towards a coordinated strategy for economic localisation*. Can be downloaded from www.regionalprogress.org.

3. Norberg-Hodge, H. (2000) *Ancient Futures: learning from Ladakh*, Rider Books.

4. Ekins, P. (1989) *The Living Economy: New Economics in the Making*, Routledge.

5. Sale, K. (1980) *Human Scale*, Coward, McCann and Geoghegan.

6. Monbiot, G. (2003) 'I Was Wrong About Trade', *The Guardian*, 24th June 2003.

7. Vandana Shiva, speaking at the Soil Association conference in Cardiff, January 2007. www.soilassociation.org/web/sa/saweb.nsf/(UNID)/6BB05A7D57DAF6A28025727200416A64?OpenDocument.

8. Korten, D. C. (2006) *The Great Turning: From Empire to Earth Community*, Berrett-Koehler Publishers.

9. Bird, J. (2007) 'Energy Security in the UK: an IPPC Factfile'. Institute for Public Policy Research. www.ippr.org.uk/ publicationsandreports/publication.asp?id=555.

10. Department for Transport (2006) *Transport Statistics Great Britain: 2006 Edition*, The Stationery Office, London.

11. Daly, H. (1993) 'The Perils of Free Trade', *Scientific American*, November 1993.

12. Quoted in Hawkes, P. (2007) *Blessed Unrest: How the Largest Movement in the World Came into Being, and Why No One Saw it Coming*, Viking.

13. Department of Transport (2006) *Road Freight Statistics*, DoT June 2006. Retrieved from www.dft.gov.uk/162259/162469/221412/221522/2 22944/coll_roadfreightstatistics2005in/roadfreigh tstatistics2005int5135.

14. Monbiot, G. (2006) *Heat: How to Stop the Planet Burning*, George Unwin Press.

15. Strahan, D. (2007) *The Last Oil Shock: A Survival Guide to the Imminent Extinction of Petroleum Man*, John Murray.

16. Mellanby, K. (1975) *Can Britain Feed Itself?*, Merlin Press.

17. For non-UK readers, MEP stands for Member of the European Parliament.

18. Lucas, C., Jones, A. & Hines, C. (2006). 'Fuelling a Food Crisis'. Available to download from www.carolinelucasmep.org.uk

19. See for example Romm, J. (2005) *The Hype About Hydrogen: Fact and Fiction in the Race to Save the Climate*. Island Press.

20. Strahan, D. (2007) *The Last Oil Shock: A Survival Guide to the Imminent Extinction of Petroleum Man*. London, John Murray.

21. Kunstler, J.H. (2005) *The Long Emergency: Surviving the Converging Catastrophes of the 21st Century*. Atlantic Monthly Press.

22. Energy Watch Group (2007) *Coal: Resources and Future Production*, EWG. www.energywatchgroup.org/files/coalreport.pdf.

23. Monbiot, G. (2007) *The New Coal Age*. www.monbiot.com, 9th October 2007. www.monbiot.com/archives/2007/10/09/

the-new-coal-age/.

24. Leggett, J. (2005) *Half Gone: Oil, Gas, Hot Air and the Global Energy Crisis*, Portobello Books.

25. Monbiot, G. (2007) 'A Lethal Solution', *The Guardian*, 27th March 2007.

26. Department of Trade and Industry (2006) *Energy Trends June 2006*. Available from www.dti.gov.uk/files/file30881.pdf. Also of note here is Chris Vernon's commentary on this, *UK Energy Trends, Coal*. Posted on The Oil Drum Europe, www.theoildrum.com/story/2006/7/5/ 15249/17646.

27. For a chilling analysis of this, see Vidal, J. (2001) 'The Looming Food Crisis', *The Guardian* G2 section, Wednesday 29th August 2007, pp.5-9.

28. Chefurka, P. (2007) 'Mexico: peak oil in action'. www.paulchefurka.ca/ Mexico%20and%20the%20Problematique.html.

29. Girling, R. (2007) 'Goodbye Beautiful Britain', *Sunday Times*, August 26th 2007.

30. See, for example, DEFRA (2007) 'Hilary Benn announces action to monitor impact of EU Agriculture Council's decision on 0% set aside rate for 2008'. DEFRA press release, 26th September 2007. www.defra.gov.uk/news/2007/ 070926b.htm.

31. Fleming, D. (2005) *Energy and the Common Purpose: descending the energy staircase with Tradable Energy Quotas*, Lean Connection Press, London.

32. Department of Environment, Food and Rural Affairs (2002) *Achieving a Better Quality of Life: review of progress towards sustainable development*.

Part Two & Chapter 5: How peak oil and climate change affect us

1. Snyder, G. (1974) 'Four Changes' in *Turtle Island*, New Directions Publications, p.101.

2. Trainer, T. (2007) *Renewable Energy Cannot Sustain a Consumer Society*, Springer Verlag.

3. Response to a question to the panel session at the Soil Association conference in Cardiff, January 2007.

4. Heinberg, R. (2007) *Peak Everything: Waking Up to the Century of Declines*, New Society Publishing.

5. Heinberg, R. (2004) *Powerdown: options and actions for a post-carbon world*, Clairview Books.

6. A permaculture teacher at Kinsale FEC, who publishes www.Zone5.org.

Chapter 6: Understanding the psychology of change

1. DiClemente, C. (2006) *Addiction and Change: how addictions develop and addicted people recover*, Guilford Publications.

2. Johnstone, C. (2006) *Find Your Power: Boost Your Inner Strengths, Break Through Blocks and Achieve Inspired Action*, Nicholas Brealey Publishing.

3. www.greatturningtimes.org.

4. See, for example, Miller, W. R. & Rollnick, S. (2002) *Motivational Interviewing: Preparing People for Change* (2nd Edition), Guilford Press.

5. Bush famously made this remark in his 2006 State of the Union address; www.whitehouse.gov/news/ releases/2006/01/20060131-10.html.

6. Miller, W.R. & Sanchez, V.C. (1993) 'Motivating young adults for treatment and lifestyle change'. In: Howard, G. (ed.) *Issues in Alcohol Use and Misuse by young adults*, University of Notre Dame Press.

7. See the Resources section.

8. Explored in more detail in James, S & Lahti, T. (2004) *The Natural Step for Communities: how cities and towns can change to sustainable practices*, New Society Publishers.

Chapter 7: Harnessing the power of a positive vision

1. Atlee, T. (2003) *The Tao of Democracy: Using Co-Intelligence to Create a World That Works For All*, The Writers' Collective.

2. www.southernsolar.co.uk/.

3. www.transitionculture.org/?p=318.

4. www.lewes.gov.uk/environment/8261.asp.

5. www.zerowaste.co.nz/.

6. www.politics.guardian.co.uk/economics/story/ 0,,1965497,00.html.

7. www.en.wikipedia.org/wiki/

Vauban,_Freiburg.

8. www.en.wikipedia.org/wiki/ Gross_national_happiness.

9. www.business.scotsman.com/ topics.cfm?tid=605&id=1868722006.

10. www.furniturenow.org.uk.

11. www.stroudcommunityagriculture.org/ promoting-CSA.php.

12. www.bbc.co.uk/dna/h2g2/A2263529.

13. www.cyclestations.co.uk.

14. www.carplus.org.uk/.

15. www.en.wikipedia.org/wiki/Appreciative_inquiry.

16. Chris is also author of Johnstone, C. (2006) *Find Your Power: Boost your inner strengths, break through blocks and achieve inspired action*, Nicholas Brealey Publishing.

17. Atlee, T. (2003) *The Tao of Democracy: Using Co-Intelligence to Create a World That Works For All*, The Writers' Collective.

18. Taken from 'Exclusive to Transition Culture! Peter Russell on life after oil, change and consciousness', Transition Culture.org, 5th February 2007.

19. Lovelock, J. (2007) *The Revenge of Gaia: Why the Earth Is Fighting Back, and How We Can Still Save Humanity*. Penguin Books.

20. Harding, S. (2006) *Animate Earth: Science, Intuition and Gaia*, Green Books.

21. Taken from 'Stephan Harding on Peak Oil', Transition Culture.org, 14th June 2006.

22. Goodwin, B. (2007) *Nature's Due: Healing Our Fragmented Culture*, Floris Books.

23. Taken from 'Brian Goodwin on Peak Oil: An Interview', Transition Culture.org, 8th May 2006.

24. Taken from 'Exclusive to Transition Culture. Fritjof Capra on Peak Oil: an Interview', Transition Culture.org, 10th May 2006.

25. Taken from 'Meg Wheatley on Energy Descent Planning : an Interview'. Transition Culture.org, 13th September 2006.

26. Taken from 'Exclusive to Transition Culture: An Interview with Tony Juniper. Part 1. Peak oil, climate change and the role of local communities', Transition Culture.org, 23rd February 2007.

27. Taken from 'An Interview with Dennis Meadows, co-author of *Limits to Growth*', Transition Culture.org, 18th September 2006.

Chapter 8: A vision for 2030

1. As proposed in Lucas, C., A. Jones & C. Hines, 'Fuelling a Food Crisis: the impact of peak oil on food security', Green Party/European Free Alliance. www.carolinelucasmep.org.uk/ framesets/publications.html.

2. See for example the Badgersett Research Corporation at www.badgersett.com and the Agroforestry Research Trust at www.agroforestry.co.uk.

3. Cart Horse Machinery, www.carthorsemachinery.com.

4. Natural Building Technologies, www.naturalbuildingproducts.co.uk.

5. S & S Cob Blocks, www.cobblocks.co.uk.

6. Shearer, P. (2006) 'The Weetabix house: This eco-charity's new HQ is packed with bales of natural fibre', *The Times*, 10th November 2006. www.property.timesonline.co.uk/article/0,,14049-2443927,00.html.

7. The Humungus Fungus Project, www.jac-by-the-stowl.co.uk.

8. Fruiting Bodies, www.fruiting-bodies.co.uk.

9. Bioregional Development Group, www.bioregional.com/programme_projects/ pap_fibres_prog/hemp%20textiles/ hemp_reports.htm#project.

10. Wood Energy Limited, www.woodenergyltd.co.uk/ft_pellets.html.

11. Healion, K. (2005) 'Case Study: Camphill Community Ballytobin, Co. Kilkenny, Ireland'. A Report for Sustainable Energy Ireland. www.sei.ie/uploadedfiles/RenewableEnergy/Task 29CamphilCommunityBallytobin.pdf.

12. Bryson, L, Jones, A, Beedle, L & York, A. 'The Glebelands-Unicorn Model: A Cooperative approach to sustainable urban food supply'. www.unicorn-grocery.co.uk/ glebelands_model.php.

13. Viljoen, A.(ed) (2005). *CPULs: Continuous Productive Urban Landscapes*, Architectural Press.

14. A formula set out in the Centre for Alternative Technology (2007) *Zero Carbon Britain*, CAT Publications.

15. Fleming, D. (2005) *Energy and the Common Purpose: descending the Energy Staircase with Tradable Energy Quotas (TEQs)*, The Lean Economy Connection.

16. Chiappini, R. (2006) *Christo and Jeanne-Claude: Revealing an Object by Concealing It*, Skira Editore.

17. Set out in detail in Greenpeace (2005) 'Decentralising Power: An Energy Revolution for the 21st Century'. Available to download from www.greenpeace.org.uk.

18. Jones, A (2002) 'Woking: Energy Services, DG and Fuel Cells for the New Millenium', *Cogeneration and On-site Power Production*, Volume 3, Issue 3, May-June 2002.

19. Such a thing already exists, the Electrisave. I have one, and it is wonderful. They have just been rebranded as the Owl wireless energy monitor. www.electricity-monitor.com.

20. McLeod, R. (2007) 'Passiv Haus, Local House: an integrated approach to zero carbon housing', MSc Architecture AEES Thesis, University of East London and CAT.

21. A concept I originally explored in an article in *Resurgence Magazine* called 'Building Miles'. Available to download from www.resurgence.org/ 2006/hopkins236.htm.

22. A concept explored in Vale, R. & Vale, B. (2002) *The New Autonomous House: Design and Planning for Sustainability*, Thames & Hudson.

23. For a deeper exploration of this concept, see Fairlie, S. (1996) *Low Impact Development: Planning and People in a Sustainable Countryside*, Jon Carpenter Publishing; and Hopkins, R. (1996) 'Permaculture: A New Approach for Rural Planning?', BSc Dissertation at the University of the West of England. The main point of contact for all of this is Chapter 7, at www.tlio.org.uk/chapter7/index.html.

24. The Land is Ours (1999) *Defining Rural Sustainability: Fifteen Criteria for Sustainable Developments in the Countryside together with Three Model Policies for Local Plans*. Available from the Chapter 7 office: The Potato Store, Flax Drayton Farm, South Petherton, Somerset, TA13, U.K. or at www.tlio.org.uk/chapter7/ defining.html.

25. These exercises were developed by Chris Salisbury of Wildwise, an excellent local environmental education organisation. www.wildwise.co.uk.

26. This isn't a mad idea I came up with – it already happens in Sweden. Have a look at an excellent report on the subject, Schonning, C. (2001) *Urine Diversion: hygienic risks and microbial guidelines*. Available as a pdf. at www.who.int/water_sanitation_health/ wastewater/urineguidelines.pdf.

Chapter 9: Kinsale – a first attempt at community visioning

1. This course is still running – check out www.kinsalefurthered.ie.

2. James, S. & Lahti, T. (2004) *Natural Step for Communities: How Cities and Towns Can Change to Sustainable Practices*, New Society Publishing.

3. You can read more about Open Space Technology in Tools for Transition No. 10, p.168.

4. Available as a free download at www.transitionculture.org.

5. You can hear an interview with Richard recorded on the last day of the Fuelling the Future conference which captures a taste of the event at www.globalpublicmedia.com/ richard_heinberg_on_the_lifeboat_radio_show.

6. You can hear the audio files of the lectures from the conference at www.fuellingthefuture.org/audio.htm.

7. An excellent introduction to the Village Building Convergence can be found at www.cityrepair.org/wiki.php/, where you will find a great short film about VBC.

8. Bill Mollison, in *Permaculture: a Designer's Manual* (Tagari Publications 1988) defines swales as "long level excavations, which can vary greatly in width and treatment from small ridges in gardens, rock piles across slopes, or deliberately-excavated hollows in flatlands and low-slope landscapes". They are used to intercept, stop and infiltrate overland water flow.

9. An interview with Catherine from the early stages of Transition Town Kinsale can be heard on Global Public Media, at www.globalpublicmedia.com/ catherine_dunne_transition_design.

10. For more information see www.willitseconomiclocalization.org.

11. The most recent interview at Global Public Media about TTK was with Klaus Harvey in October 2007, and can be heard at www.globalpublicmedia.com/ kinsale_two_years_on.

12. A piece I wrote on Transition Culture in 2006, 6 months after the publication of the Kinsale EDAP, gave a flavour of the breadth of its dissemination. www.transitionculture.org/ 2006/02/13/kinsale-action-plan-sending-up-shoots-around-the-world.

13. Taken from Joel Barker's DVD *The Power of Vision*. Available from www.joelbarkertrainingvideos.com/ The_Power_Of_Vision.htm#Purchase.

Part Three & Chapter 10: The Transition concept

1. From a Lecture on Public Policy at the University of Cambridge entitled 'The transition economy: a future beyond oil' (while he was still Environment Minister), 5th March 2007. Available at www.defra.gov.uk/corporate/ ministers/speeches/david-miliband/ dm070305.htm.

2. Shepard, H. A. (1974) 'Rules of Thumb for Change Agenda', reprinted at www.nickheap.co.uk.

3. McNamara, A. (2007) 'New Queensland Sustainability Minister on the future with less oil'. An interview with Andi Hazelwood of Global Public Media. www.globalpublicmedia.com/ new_queensland_sustainability_minister_on_the_f uture_with_less_oil.

4. Holmgren, D. (2003) 'What Is Sustainability?' from Sustainability Network Update 31E, 9th September 2003, CSIRO Sustainability Network. Available at www.bml.csiro.au/susnetnl/ netwl31Eb.pdf.

5. Initially set out in Mollison, B. & Holmgren, D. (1978) *Permaculture One: A Perennial Agriculture for Human Settlements*, Tagari Publications, but best defined in Holmgren, D. (2007) *Permaculture: Principles and Pathways beyond Sustainability*, Holmgren Design Services.

6. To find out about you can take a permaculture course in the UK, visit the website of the Permaculture Association (Britain) at www.permaculture.org.uk. Courses run throughout the UK and in a variety of formats. A permaculture course is a life-changing experience, and, I would argue, essential for this work.

7. Mollison, B. (1988) *Permaculture: A Designers Manual*, Tagari Publications.

8. Stewart, E. 'A Second Challenge to the Movement', *Permaculture Activist* No. 57.

9. Seager, A. (2007) 'Steep decline in oil production brings risk of war and unrest, says new study', *The Guardian*, 22nd October 2007.

10. Adam, D. (2007) 'Carbon output rising faster than forecast, says study', *The Guardian*, 23rd October 2007.

11. These issues are explored in far more depth and incomparably well in Joanna Macy (1999) *Coming Back to Life: Practices to Reconnect Our Lives, Our World*, New Society Publishing.

12. You can obtain DVD copies and find out more at www.whatawaytogomovie.com. An excellent film exploring the 'heart' aspects of energy descent. Powerful, moving and thought provoking, I found it extremely insightful, but not one for beginners.

13. The Gaia Foundation of Western Australia's website is www.gaia.iinet.net.au.

14. See, for example, Sale, K. (1980) *Human Scale*, Coward, McCann & Geoghegan; and Morris, D. & Hess, K. (1975) *Neighbourhood Power: The New Localism*, Beacon Press.

15. www.transitioncitybristol.org.

16. Winter, M. (2006) *Peak Oil Prep: three things you can do to prepare for peak oil, climate change and economic collapse*, Westsong Publishing.

17. An excellent overview of this municipal-scale work can be found in Lerch, D. (2007) *Post Carbon Cities: Planning for Energy and Climate Uncertainty. A guidebook on peak oil and global warming for local governments*, Post Carbon Institute. Can be purchased from www.postcarboncities.net.

Chapter 11: How to start a Transition Initiative

1. The best place to start finding out about Open Space on the web is www.openspaceworld.org.

2. Owen, H. (1993) *Open Space Technology: A Users Guide*, Berrett-Koehler.

3. For a list of Open Space practitioners in the UK, see www.openspaceuk.com/index.php/practictioners, or contact the Transition Network.

4. This can be downloaded at www.transitionculture.org.

5. More information about the Fishbowl technique can be found on Wikipedia, at www.en.wikipedia.org/wiki/Fishbowl_(conversation).

6. The Transition Network is currently developing a template for this, as well as a suggested 'timeline' around which events can be organised.

Chapter 12: The first year of Transition Town Totnes

1. You can see the notes generated during the day at www.transitiontowns.org/Totnes/.

2. From www.theworldcafe.com.

3. www.secondnature.ie.

4. www.theworldcafe.com. 'From Café Principles in Action'. www.theworldcafe.com/know-how.htm#explore.

5. Apart from www.theworldcafe.com, the other key references on World Café are; World Café Community (2002) *Café to Go! A quick reference guide for putting conversations to work*, Whole Systems Associates, available from www.theworldcafe.com/articles/cafetogo.pdf; and Brown, J., Isaacs, D. & The World Cafe Community (2005) *The World Cafe: Shaping Our Futures Through Conversations That Matter*, Berrett-Koehler Publishing. Some other resources can be found at www.theworldcafe.com/hosting.htm.

6. www.movingsounds.org.

7. You can see the film that Keith made for Transition Town Lewes as a promotion for their Unleashing on YouTube. It was shown by his local cinema as a trailer in the two months running up to their Unleashing.

8. Agroforestry Research Trust, 46 Hunters Moon, Dartington, Totnes, Devon TQ9 6JT. www.agroforestry.co.uk.

9. McIntosh, A. (2004) *Soil and Soul: People Versus Corporate Power*, Aurum Press.

10. You can hear the Radio Scotland piece at www.transitiontowns.org/TransitionNetwork/TransitionNetwork.

11. University of Liverpool Oil Depletion Impact Group. www.liv.ac.uk/managementschool/odig/index.htm.

12. You can read a more detailed overview of the material covered on the course at www.transitionculture.org/skilling-up-for-powerdown-course-notes/.

13. www.agroforestry.co.uk.

14. www.berkshares.org.

15. www.saltspringtoday.com.

16. You can see this film at www.youtube.com/watch?v=aUuC_YecrfE.

17. A publication we found extremely useful in designing the Totnes Pound was Lietaer, B. & Hallsmith, G. (2006) *Community Currency Guide*, Global Community Initiatives. Available as a free download from www.lyttelton.net.nz/documents/timebank/community_currency.pdf.

Chapter 13: The viral spread of the Transition concept

1. You can hear the podcasts from the conference at www.soilassociation.org/conference and read the booklet that was produced for the event at www.transitionculture.org/2007/01/30/one-planet-agriculture-the-case-for-action-download-the-booklet/.

2. To follow its progress, visit www.lampeter.org/english/tt/index.html.

3. Which you can hear online at www.transitiontowns.org/TransitionNetwork/TransitionNetwork.

4. Somewhat akin to www.bookcrossing.com.

5. You can see a short film of Duncan talking about Transition Brixton at www.transitionculture.org/2007/08/30/duncan-law/.

6. Explored in more detail in Thompson, S., Abdullah, S., Marks, N., Simms, A. & Johnson, V. (2007) *The European unHappy Planet Index: an index of carbon efficiency and well being in the EU*, New Economics Foundation, London.

7. www.news.bbc.co.uk/1/hi/magazine/6676967.stm.

8. Statistics taken from Boyle, D., Cordon, C., Potts, R. & Simms, A. (2007) *Are You Happy?*, New Economics Foundation.

Resources

DVDs

The End of Suburbia: Oil Depletion and the Collapse of The American Dream by The Electric Wallpaper Co. www.endofsuburbia.com.

The Power of Community: How Cuba Survived Peak Oil by Community Solution. www.powerofcommunity.org.

A Crude Awakening: The Oil Crash by Lava Productions AG. www.oilcrashmovie.com.

Crude Impact by Vista Clara Films. www.vistaclarafilms.com.

Peak Oil: Imposed by Nature by Tropos Dokumentar. www.troposdoc.com.

Escape from Suburbia: Beyond the American Dream. www.escapefromsuburbia.com

What a Way to Go: Life at the End of Empire by Tim Bennett/VisionQuest Pictures. www.whatawaytogomovie.com.

An Inconvenient Truth by Paramount Classics and Participant Productions. www.aninconvenienttruth.co.uk.

Online videos

Real Oil Crisis by ABC (the Australian Broadcasting Corporation). www.abc.net.au/catalyst/stories/s1515141.htm

Our Oil Addiction by WCCO-TV. www.wcco.com/energy

Peak Oil by ABC. www.abc.net.au/4corners/special_eds/20060710/default_full.htm.

Global Public Media is an essential online archive resource of interviews and film about peak oil and solutions.

Crude: the incredible journey of oil. ABC Science. www.abc.net.au/science/crude/.

Five essential books to read when you've finished this one

Heinberg, R. (2007) *Peak Everything: Waking Up to the Century of Declines*, New Society Publishing.

Holmgren, D. (2003) *Permaculture: Principles and Pathways Beyond Sustainability*, Holmgren Design Services.

Homer-Dixon, T. (2003) *The Upside of Down: Catastrophe, Creativity and the Renewal of Civilisation*, Souvenir Press.

Strahan, D. (2007) *The Last Oil Shock: A Survival Guide to the Imminent Extinction of Petroleum Man*, John Murray Publishing.

Trainer, T. (2007) *Renewable Energy Cannot Sustain a Consumer Society*, Springer Verlag.

Peak oil

Bentley, R.W. (2002) 'Global oil & gas depletion: an overview', *Energy Policy* 30, pp.189–205.

Brown, L.R. (2006a) *Plan B. 2.0: Rescuing a Planet Under Stress and a Civilization in Trouble*, W.W. Norton & Co Ltd.

Campbell, C.J. and Laherrere, J. (1998). 'The End of Cheap Oil', *Scientific American*, Vol. 278, No. 3. pp.78-83.

Campbell, C.J. (2005a) *Oil Crisis*, Multi-Science Publishing Co. Ltd.

Deffeyes, K.S. (2001) *Hubbert's Peak: The Impending World Oil Shortage*, Princeton University Press.

Deffeyes, K.S. (2005) *Beyond Oil : The View from Hubbert's Peak*, Hill and Wang.

Hartmann, T. (1999) *The Last Hours of Ancient Sunlight: waking up to personal and global transformation*, Hodder & Stoughton.

Heinberg, R. (2003) *The Party's Over: Oil, War and the Fate of Industrial Societies*, New Society.

Heinberg, R. (2004) *Powerdown: Options and Actions for a Post-Carbon World*, New Society.

Heinberg, R. (2006) *The Oil Depletion Protocol: A Plan to Avert Oil Wars, Terrorism and Economic Collapse*, Clairview Books.

Hirsch, R.L., Bezdek, R. Wendling, R. (2005a.) *Peaking of World Oil Production: Impacts, Mitigation and Risk Management*, National Energy Technology Laboratory, US Department of Energy.

Hirsch, R.L. (2005b.) 'Supply and Demand. Shaping the peak of world oil production', *World Oil*, October 2005.

Howe, J.G. (2003) *The End of Fossil Energy and the Last Chance for Sustainability*, McIntire Publishing.

Klare, M.T. (2004) *Blood and Oil: The Dangers and Consequences of America's Growing Dependency on Imported Petroleum* (The American Empire Project), Metropolitan Books.

Kraemer, T.D. (2006) *Addicted to Oil: Strategic Implications of American Oil Policy*, Strategic Studies Institute of the US Army War College.

Kunstler, J.H. (2005) *The Long Emergency: Surviving the Converging Catastrophes of the 21st Century*, Atlantic Monthly Press.

Leggett, J. (2005b.) *Half Gone: oil, gas, hot air and the global energy crisis*, Portobello Books.

Mobbs, P. (2005) *Energy Beyond Oil*, Matador Books.

Simmonds, M.R. (2005) *Twilight in the Desert: The Coming Saudi Oil Shock and the World Economy*, John Wiley.

Tertzakian, P. (2006) *A Thousand Barrels a Second: The Coming Oil Break Point and The Challenges Facing An Energy Dependent World*, McGraw-Hill.

Staniford, S. (2006) *Why peak oil is probably about now*. The Oil Drum.com. www.theoildrum.com/story/2006/3/1/3402/63420

Climate change

Dow, K. and Downing, T.A. (2006) *The Atlas of Climate Change*, Earthscan Books.

Hillman, M. (2005). *How to Save The Planet*, Penguin Books.

Goodall, C. (2007) *How to Live a Low-Carbon Life. The Individual's Guide to Stopping Climate Change*, Earthscan.

Henson, R. (2006) *The Rough Guide to Climate Change: the symptoms, the science, the solutions*, Rough Guides, London.

I-Count (2006) *Your step-by-step guide to Climate Bliss*, Penguin Books.

Lovelock, J. (2007) *The Revenge of Gaia: Why the Earth Is Fighting Back, and How We Can Still Save Humanity*, Penguin Books.

Lynas, M. (2007) *Six Degrees: our future on a hotter planet*, Fourth Estate.

Monbiot, G. (2007) *Heat: how to stop the planet burning*, Penguin Books.

Pearce, F. (2007) *Last Generation: How Nature Will Take Her Revenge for Climate Change*, Eden Books/Transworld.

Simms, A, Kjell, P. & Woodland, D. (2005) *Mirage and Oasis: energy choices in an age of global warming*, New Economics Foundation. Free download from www.neweconomics.org.

Spratt, D. (2007) *The Big Melt: lessons from the Arctic Summer of 2007*, Carbon Equity. Available from www.carbonequity.info.

Energy descent / Transitions

Abdullah, S. (1999) *Creating a World That Works For All*, Berrett-Koehler.

Bates, A. (2007) *The Post-Petroleum Survival Guide and Cookbook: Recipes for Changing Times*. New Society Publishers.

Brand, S. (1999) *The Clock of the Long Now: Time and Responsibility*, Phoenix Press.

Centre for Alternative Technology (2007) *Zero Carbon Britain*, CAT Publications, Machynlleth.

Gladwell, M. (2000) *The Tipping Point: how little things can make a big difference*, Abacus.

Haggis, G. (2007) *The Energy Challenge: Finding Solutions to the Problems of Global Warming and Future Energy Supply*, Matador, Leicester.

Holmgren, D. (2004) *Permaculture: Principles and Pathways Beyond Sustainability*, Holmgren Design Press.

Hopkins, R. (2007) *Energy Descent Pathways: evaluating potential responses to peak oil*. An MSc dissertation for the University of Plymouth. Available from www.transitionculture.org.

Korten, D.C. (2006) *The Great Turning: from Empire to Earth Community*, Berett-Koehler Publishers.

Korten, D.C. (2000) *The Post Corporate World: Life After Capitalism*, Berrett-Koehler Publishers.

Laszlo, E. (1994) *The Choice: Evolution or Extinction? A Thinking Person's Guide to Global Issues*, Tarcher/Putnam.

Laszlo, E. (2006) *The Chaos Point: the world at the crossroads*, Piatkus Press.

Odum, H.T. & Odum, E.C. (2001) *A Prosperous Way Down: principles and policies*, University Press of Colorado.

Solnit, R. (2004) *Hope in the Dark: the Untold History of People Power*, Canongate.

Trainer, T. (2007) *Renewable Energy Cannot Sustain a Consumer Society*, Springer Verlag.

Walker, B & Salt, D. (2006) *Resilience Thinking: sustaining ecosystems and people in a changing world*, Island Press.

Winter, M. (2006) *Peak Oil Prep: three things you can do to prepare for peak oil, climate change and economic collapse*, Westsong Publishing.

Permaculture design

Holmgren, D. (2004) *Permaculture: Principles and Pathways Beyond Sustainability*, Holmgren Design Press. Quite simply the best book of the past 15 years.

Mollison, B. (1988) *Permaculture: a Designers Manual*, Tagari Press.

Whitefield, P. (2000) *Permaculture in a Nutshell*, Permanent Publications.

Whitefield, P. (2005) *The Earth Care Manual*, Permanent Publications.

Also highly recommended is a subscription to *Permaculture Magazine* (www.permaculture.co.uk) and to *The Permaculture Activist* (www.permacultureactvist.net).

Food, gardening and growing

Ableman, M. (1998) *On Good Land: The Autobiography of an Urban Farm*, Chronicle Books.

Bartholemew, M. (2006) *All New Square Foot Gardening: Grow More in Less Space!*, Cool Springs Press.

Fern, K. (1997) *Plants for a Future: Edible and Useful Plants for a Healthier World*, Permanent Publications.

Jeavons, J. (2005) *How to grow more vegetables than you ever thought possible on less land than you can imagine*, Ten Speed Press.

Whitefield, P. (1996) *How to Make a Forest Garden*, Permanent Publications.

Larkcom, J. (1998) *Grow Your Own*, Frances Lincoln.

Guerra, M. (2005) *The Edible Container Garden: Fresh Food from Tiny Spaces*, Gaia Books.

Hickmott, S. (2003) *Growing Unusual Vegetables: weird and wonderful vegetables and how to grow them*, Ecologic Books.

Lucas, C, Jones, A.& Hines, C. (2007) *Fuelling a Food Crisis: the impact of peak oil on food security*. A free download from www.carolinelucasmep.org.uk/publications/pdfs_and_word/Fuelling%20a%20food%20crisis%20FINAL%20Dec06.pdf

Mellanby, K. (1975) *Can Britain Feed Itself?*, Merlin Press.

Pullen, M. (2004) *Valuable Vegetables: Growing for Pleasure and Profit*, Ecologic Books.

Tudge, C. (2003) *So Shall We Reap: What's Gone Wrong With the World's Food: And How To Fix It*, Penguin.

Natural building and insights for post-peak construction

Alexander, C. (2002-2005) *The Nature of Order: An Essay on the Art of Building and the Nature of the Universe (4 volume series)*, Centre for Environmental Structure, Berkeley, California.

Alexander, C. et al (1977) *A Pattern Language*, Oxford University Press.

Borer, P & Harris, C. (2000) *The Whole House Book*, CAT Publications. Machynlleth.

Clifton-Taylor, A. (1987) *The Pattern of English Building (4th edition)*, Faber and Faber.

Day, C. (2003) *Consensus Design: socially inclusive process*, Architectural Press.

Day, C. (2004) *Places of the Soul: Architecture and Environmental Design as a Healing Art*, Architectural Press.

Evans, I, Smith, M & Smiley, L. (2002) *Hand Sculpted House: a practical and philosophical guide to building a cob cottage*. Chelsea Green.

Kennedy, J., Smith, M. and Wanek, C. eds. (2002). *The Art of Natural Building: Design, Construction, Resources*, New Society Publishers

Magwood, C & Mack, P. (2000) *Straw Bale Building: how to plan, design and build with straw*. Gabriola Island, New Society Publishing.

Roaf, S, Crichton, D. & Nicol, F. (2005) *Adapting Buildings and Cities for Climate Change: A 21st Century Survival Guide*. Architectural Press.

Roy, R. (2003) *Cordwood Building: state of the art*, New Society Publishing.

Woolley, T. (2006) *Natural Building: A Guide to Materials and Techniques*, The Crowood Press.

Thinking approaches, tools and change processes

Atlee, T. (2003) *The Tao of Democracy. Using Co-Intelligence to create a world that works for all*, The Writers' Collective.

Holman, P & Devane, T. (1999) *The Change Handbook: Group Methods for Shaping the Future*, Berrett-Koehler.

James, S. & Lahti, T. (2004) *The Natural Step for Communities: how cities and towns can change to sustainable practices*, New Society Publishing.

Johnstone, C. (2006) *Find Your Power: Boost your inner strengths, break through blocks and achieve inspired action*, Nicholas Brealey Publishing.

Macy, J. & Brown, M.Y. (1998) *Coming Back to Life: Practices to Reconnect Out Lives, Our World*, New Society Publishers.

Macy, J. (2000) *Widening Circles: A Memoir*. Gabriola Island, New Society Publishers.

McKenzie-Mohr, D. & Smith, W. (1999) *Fostering Sustainable Behaviour: An introduction to community-based social marketing*, New Society.

Owen, H. (1993) *Open Space Technology: A Users Guide*, Berrett-Koehler.

Seligman, M.E.P. (2006) *Learned Optimism: how to change your mind and your life*, Vintage.

Trapese Collective (2007) *Do It Yourself: A Handbook For Changing Our World*, Pluto Press.

Westley, F, Zimmerman, B. & Patton, M.Q. (2007) *Getting to Maybe: How the World is Changed*, Vintage.

World Café Community (2002) *Café to Go! A quick reference guide for putting conversations to work*, Whole Systems Associates.

Zerubavel, E. (2006) *The Elephant in the Room: Silence and Denial in Everyday Life*, Oxford University Press.

Insights on resilience and its historical disappearance

Berger, J. (1992) *Pig Earth*, Chatto & Windus.

Callandar, R. (2000) *History in Birse: Volumes 1-4*, Birse Community Trust.

Collis, J.S. (1973) *The Worm Forgives The Plough*, Penguin.

Couto, D. (2002) 'How Resilience Works', *Harvard Business Review* Vol. 80, No. 5. May 2002.

Dartington Rural Archive (1985) *Horsepower*, Spindle Press.

Drummond, J.C. & Wilbraham, A. (1958) *The Englishman's Food: A History of 5 Centuries of English Diet*, Readers Union/Jonathan Cape.

Fleming, D. (2007) *Lean Logic: The Dictionary of Environmental Manners*. Forthcoming.

Gardiner, J. (2004) *Wartime Britain 1939-45*, Headline Book Publishing.

Hammond, R.J. (1954) *Food and Agriculture in Britain 1939-45. Aspects of Wartime Control*, Stanford University Press.

Homer-Dixon, T. (2003) *The Upside of Down: Catastrophe, Creativity and the Renewal of Civilisation*, Souvenir Press.

Hyams, E. (1952) *Soil and Civilisation*, Thames and Hudson.

Jones, S.R. (1936) *English Village Homes*, Batsford.

Newby, H. (1980) *Green and Pleasant Land?*

Social Change in Rural England, Pelican Books.

Norberg-Hodge, H. (2000) *Ancient Futures: Learning from Ladakh*, Rider Publishing.

Ponting, C. (1991) *A Green History of the World; The Environment and the Collapse of Great Civilisations*, Penguin.

Rackham, O. (1995) *Trees and Woodlands in the British Landscape. The Complete History of Britain's Trees, Woods and Hedgerows*, Weidenfeld & Nicolson.

Reader, J. (1988) *Man on Earth*, Penguin Books.

Sturt, G. (1963) *The Wheelwrights Shop*, Cambridge University Press.

Viljoen, A. (2006) *CPULs: Continuous Productive Urban Landscapes*, Architectural Press.

Walker, B & Salt, D. (2006) *Resilience Thinking: Sustaining Ecosystems and People in a Changing World*. Washington, Island Press.

Wilkinson, R.G. (1973) *Poverty and Progress*, Cambridge University Press.

Zweiniger-Bargielowska, I. (2000) *Austerity in Britain: rationing, controls, and consumption 1939-1955*, Oxford University Press.

Insights from addiction

DiClemente, C.C. (2003) *Addiction and Change: how addictions develop and addicted people recover*, Guilford Press.

Glendinning, C. (1994) *My Name is Chellis and I'm In Recovery from Western Civilisation*, Shambhala Publishing.

La Chance, A. (1991) *Greenspirit: Twelve Steps in Ecological Spirituality – an individual, cultural and planetary therapy*, Element Books.

Miller, W.R. & Rollnick, S. (2002) *Motivational Interviewing: preparing people for change*, Guilford Press.

Schaef, A.W. (1987) *When Society Becomes an Addict*, Harper Row Publishers.

Rethinking economics

Boyle, D. (2002) *The Money Changers: currency reform from Aristotle to e-cash*, Earthscan.

Cato, M.S. (2006) *Market Schmarket: building the post-capitalist economy*, New Clarion Press.

Dawson, J. (2006) *Ecovillages: New Frontiers for Sustainability*. Schumacher Briefing no. 12, Green Books.

Douthwaite, R. (1996) *Short Circuit: strengthening local economies for security in an unstable world*, Green Books.

Douthwaite, R. (1999) *The Growth Illusion: How Economic Growth Has Enriched the Few, Impoverished the Many and Endangered the Planet*, Green Books.

Fleming, D. (2005) *Energy and the Common Purpose: descending the energy staircase with Tradable Energy Quotas*, Lean Connection Press, London.

Greco, T. (2001) *Money: understanding and creating alternatives to legal tender*, Chelsea Green Publishing.

Johanisova, N. (2005) *Living in the Cracks: a look at rural social enterprises in Britain and the Czech Republic*, Green Books.

Korten, D.C. (2000) *The Post Corporate World: Life After Capitalism*. San Francisco, Berrett-Koehler Publishing.

Lietaer, B. (2001) *The Future of Money: creating new wealth, work and a wiser world*, Century.

Perkins, J. (2005) *Confessions of an Economic Hit Man*, Ebury Press/Berret-Koehler.

Scott Cato, M. (2006) *Market Schmarket: Building the Post Capitalist Economy*, New Clarion Press.

Simms, A. (2005) *Ecological Debt: The Health of the Planet and the Wealth of Nations*, Pluto Press.

Localisation

Cavanagh, J. & Mander, J. (2004) *Alternatives to Economic Globalisation: a better world is possible*, Berrett-Koehler.

Ekins, P. (1989) *Towards a New Economics: on the theory and practice of self reliance*, Routledge.

Goldsmith, E. & Mander, J. (eds) (2001) *The Case Against the Global Economy and For a Turn Towards Localisation*, Earthscan.

Greenpeace (2006) *Decentralising Power: An Energy Revolution for the 21st Century*. www.greenpeace.org.uk.

Hines, C. (2000) *Localisation: A Global Manifesto*, Earthscan.

Norberg-Hodge, H. (2000) *Ancient Futures: Learning from Ladakh*, Rider Publishing.

Norberg-Hodge, H. (2002) *Bringing the Local Economy Home: Local Alternatives to Global Agribusiness*, Zed Books.

Sale, K. (1980) *Human Scale*, Coward, McCann & Geoghegan.

Sale, K. (1985) *Dwellers in the Land: The Bioregional Vision*, Sierra Club Books.

Shuman, M. (2000) *Going Local: creating self reliant communities in a global age*, Routledge.

Woodin, M. & Lucas, C. (2004) *Green Alternatives to Globalisation: a manifesto*, Pluto Press.

Hard to categorise but wonderful nonetheless

Cockayne, E. (2007) *Hubbub: Filth, Noise and Stench in England*, Yale University Press.

Freese, B. (2006) *Coal: A Human History*, Arrow Books.

Harding, S. (2006) *Animate Earth: science, intuition and Gaia*, Green Books.

Some essential websites

www.transitionculture.org. Where the concepts in this book are explored in more depth.

www.energybulletin.net. Essential – make it the one website you visit every day.

www.poweringdown.blogspot.com. An excellent blog site exploring issues around energy descent.

www.richardheinberg.com. Richard Heinberg's site which contains the archive of his excellent Museletters.

www.theoildrum.com. Quite technical, and not really one for the novice, but an excellent source of up-to-date analysis.

www.dynamiccities.squarespace.com. The Dynamic Cities Project, which is doing some of the most essential thinking on all this.

www.postcarbon.org. Post Carbon Institute.

www.zone5.org. Permaculture and powerdown in Ireland.

www.casaubonsbook.blogspot.com. Sharon Astyk's excellent blog.

www.pathtofreedom.com. A great hands-on site.

www.livingonthecusp.org. Naresh Giangrande's website.

www.globalpublicmedia.com. An invaluable resource, films, transcripts and audio files of interviews and articles by many of the leading thinkers in this field.

www.lastoilshock.com. David Strahan's website, with some excellent interviews and articles.

www.community solution.org. The Community Solution, a wonderful organization in the US, the people who produced The Power of Community DVD, among other things.

www.powerswitch. org.uk. One of the best UK peak oil websites.

Index

Page numbers in *italic* refer to Figures and illustrations; those followed by (s) refer to sidebars.